Biotyping in the Clinical Microbiology Laboratory

Publication Number 1023

AMERICAN LECTURE SERIES®

A Monograph in

The BANNERSTONE DIVISION *of*
AMERICAN LECTURES IN CLINICAL MICROBIOLOGY

Edited by

ALBERT BALOWS, Ph.D.

Director, Bacteriology Division
Center for Disease Control
Atlanta, Georgia

Biotyping in the Clinical Microbiology Laboratory

Edited by

ALBERT BALOWS, Ph.D.

*Center for Disease Control
Atlanta, Georgia*

and

HENRY D. ISENBERG, Ph.D.

*Long Island Jewish/Hillside Medical Center
New Hyde Park, New York*

CHARLES C THOMAS • PUBLISHER
Springfield • Illinois • U.S.A.

Published and Distributed Throughout the World by
CHARLES C THOMAS • PUBLISHER
Bannerstone House
301-327 East Lawrence Avenue, Springfield, Illinois, U.S.A.

This book is protected by copyright. No part of it may be reproduced in any manner without written permission from the publisher.

© *1978, by* CHARLES C THOMAS • PUBLISHER
ISBN 0-398-03806-6
Library of Congress Catalog Card Number: 78-4865

With THOMAS BOOKS *careful attention is given to all details of manufacturing and design. It is the Publisher's desire to present books that are satisfactory as to their physical qualities and artistic possibilities and appropriate for their particular use.* THOMAS BOOKS *will be true to those laws of quality that assure a good name and good will.*

Printed in the United States of America
R-1

Library of Congress Cataloging in Publication Data
Main entry under title:

Biotyping in the clinical microbiology laboratory.

  (American lecture series; publication no. 1023)
  Bibliography: p.
  Includes index.
  1. Bacteriophage typing. 2. Bacteriology, Medical.
I. Balows, Albert. II. Isenberg, Henry D. [DNLM:
1. Microbiological technics. QW25.3 B615]
QR67.B56   616.01′4   78-4865
ISBN 0-398-03806-6

CONTRIBUTORS

STEPHEN D. ALLEN, M.D.: Indiana University Medical Center, Indianapolis, Indiana.

ANN Y. ARMFIELD: Center for Disease Control, Atlanta, Georgia.

ALBERT BALOWS, Ph.D.: Center for Disease Control, Atlanta, Georgia.

DON J. BRENNER, Ph.D.: Center for Disease Control, Atlanta, Georgia.

V.R. DOWELL, JR., Ph.D.: Center for Disease Control, Atlanta, Georgia.

J.J. FARMER III, Ph.D.: Center for Disease Control, Atlanta, Georgia.

HENRY D. ISENBERG, Ph.D.: Long Island Jewish/Hillside Medical Center, New Hyde Park, New York.

GEORGE L. LOMBARD, Dr.P.H.: Center for Disease Control, Atlanta, Georgia.

JAMES D. MacLOWRY, M.D.: Clinical Center, National Institutes of Health, Bethesda, Maryland.

MICHAEL D. STARGEL, M.D.: Parkland Memorial Hospital, Dallas, Texas.

FRANCES S. THOMPSON: Center for Disease Control, Atlanta, Georgia.

HAZEL W. WILKINSON, Ph.D.: Center for Disease Control, Atlanta, Georgia.

FOREWORD

THE GENESIS OF THIS SERIES, *The American Lecture Series in Clinical Microbiology*, stems from the concerted efforts of the Editor and the Publisher to provide a forum from which well-qualified and distinguished authors may present, either as a book or monograph, their views on any aspect of clinical microbiology. Our definition of clinical microbiology is conceived to encompass the broadest aspects of medical microbiology not only as it is applied to the clinical laboratory but equally to the research laboratory and to theoretical considerations. In the clinical microbiology laboratory we are concerned with differences in morphology, biochemical behavior and antigenic patterns as a means of microbial identification. In the research laboratory or when we employ microoganisms as a model in theoretical biology, our interest is often focused not so much on the above differences but rather on the similarities between microorganisms. However, it must be appreciated that even though there are many similarities between cells, there are important differences between major types of cells which set very definite limits on the cellular behavior. Unless this is understood it is impossible to discern common denominators.

We are also concerned with the relationships between microorganisms and disease — any microorganism and any disease. Implicit in these relations is the role of the host which forms the third arm of the triangle: microorganism, disease and host. In this series we plan to explore each of these; singly where possible for factual information and in combination for an understanding of the myriad of interrelationships that exist. This necessitates the application of basic principles of biology and may, at times, require the emergence of new theoretical concepts which will create new principles or modify existing ones. Above all, our aim is to present well-documented books which will be informative, instructive and useful, creating a sense of satisfaction to both the reader and the author.

Closely intertwined with the above *raison d'etre* is our desire to produce a series which will be read not only for the pleasure of knowledge but which will also enhance the reader's professional skill and extend his technical ability. *The American Lecture Series in Clinical Microbiology* is dedicated to biologists — be they physicians, scientists or teachers — in the hope that this series will foster better appreciation of mutual problems and help close the gap between theoretical and applied microbiology.

The application of sound, scientific microbiologic techniques in the clinical microbiology laboratory is becoming increasingly apparent as we strive to cope effectively with a wide variety of bacterial infectious diseases.

The sharp line of demarcation that once existed between those bacteria that are pathogens and those that are not is fading rapidly. The alteration of man's biosphere — particularly the microbial inhabitants of that biosphere — has created selective pressures which have given rise to bacterial species with newly acquired properties that may contribute to their pathogenicity. When one adds to this the startling observation that in many instances it is the host that determines whether or not a given organism will be able to produce disease, we must face reality and recognize the need to fully characterize a given isolate or, more properly, a given number of similar or identical isolates in order to understand better the factors that have contributed to an outbreak. This monograph attempts to place in proper perspective the need for biotyping, how it can be approached in any competent microbiology laboratory, and how the results can be utilized to resolve those questions relating to the causative agent that would otherwise go unanswered simply because the microbiologist may feel that biotyping always requires sophisticated technology. The authors were carefully selected because of the expertise and their awareness of the prevailing situation in most clinical microbiology laboratories.

The reader of this monograph should develop a keener appreciation of what biotyping can be done "at the local level" in the investigation of various types of outbreaks and in establishing a monitoring system tailor-made to meet the needs of his or her institution.

Albert Balows, Ph.D.

CONTENTS

	Page
Foreword	vii

Chapter

1. INTRODUCTION TO BIOTYPING IN CLINICAL MICROBIOLOGY — *A. Balows and H. D. Isenberg* 3
2. THE MEANING AND CLINICAL SIGNIFICANCE OF BIOTYPES — *J. D. MacLowry* 6
3. BIOTYPING OF ENTEROBACTERIACEAE IN THE CLINICAL LABORATORY — *D. J. Brenner* 12
4. BIOTYPING OF NONFERMENTATIVE, GRAM-NEGATIVE BACTERIA — *J. J. Farmer III* 26
5. STREPTOCOCCAL GROUPS OF CLINICAL SIGNIFICANCE — *H. W. Wilkinson* 33
6. BIOTYPING OF ANAEROBIC BACTERIA ASSOCIATED WITH HUMAN DISEASE — *V. R. Dowell, Jr.; G. L. Lombard; M. D. Stargel; S. D. Allen; F. S. Thompson; and A. Y. Armfield* 47
7. STANDARDIZATION AND AUTOMATION IN BIOTYPING — *J. J. Farmer III* 68
8. PERSPECTIVE OF BACTERIAL BIOTYPING FOR CLINICAL MICROBIOLOGY — *H. D. Isenberg and A. Balows* 98

Author Index 107
Subject Index 111

Biotyping in the Clinical Microbiology Laboratory

Chapter 1

INTRODUCTION TO BIOTYPING IN CLINICAL MICROBIOLOGY

A. BALOWS AND H. D. ISENBERG

"Do not fear to repeat what has already been said. Men need truth dinned into their ears, many times and from all sides. The first rumor makes them prick up their ears, the second registers and the third enters." (René Laënnec, 1781-1826).

LAËNNEC, as evidenced by this advice, certainly had his finger on the pulse of man's ability to retain what he hears and reads. The role, the need, and the importance of biotyping bacterial isolates have been presented in various publications. Still, many microbiologists, not to mention nonmicrobiologists, do not fully understand the value of biotyping in any clinical microbiology laboratory. The purpose of this monograph is to present the basic concepts of biotyping for the "third" time so that the information will "enter" and be used.

The clinical microbiology laboratory can and should be the focal point for collecting biotyping data that may indicate a potential outbreak, multiply resistant species, and add to any of several other important applied aspects of bacterial fingerprinting in the hospital or community. By and large, this monitoring can be conducted with little or no additional cost. All that is required is a good microbiology laboratory and possibly some slight revisions in record keeping. The monitoring will make it possible, for example, to flag the early appearance of an unusual biochemical reaction in *Klebsiella pneumoniae* isolates or a repetitive antibiogram in *Pseudomonas aeruginosa* isolates.

Once a nosocomial outbreak has been identified, the laboratory has an important investigative role. If necessary, the microbiologist should be prepared to collect all isolates pertaining to the outbreak and send them to a reference laboratory for more definitive typing. Biotyping has a practical aspect, but it also demonstrates the great diversities and the striking similarities that bacteria display.

To fully utilize biotyping in any microbiology laboratory, one must recall a number of often stated and frequently forgotten fundamentals:

1. Biotyping, as used in the broad sense, refers to any systematic method which will demonstrate phenotypic similarities or dissimilarities between two or more isolates that have been previously identified as belonging to the same genus and species. In some instances, biotyping strengthens the validity of the previous identification.

2. Biotyping may involve comparison of antigens, antibiograms, biochemical or substrate utilization patterns, lytic reactions to a set of bacteriophages, responses to a set of bacteriocins, qualitative or quantitative measurements of metabolic end products, or any combination of the above.

3. A valid biotyping system must be based on well-developed and well-described technology. If the results are to be useful, the reagents, apparatus, media, and methods must be standardized, and appropriate controls must be included in the tests.

4. Results obtained with a given biotyping system performed in two laboratories can be compared only when the exactness of the method used in the two laboratories has been established. Too often, this important fact is neglected and results are misinterpreted.

With appropriate biotyping data, the microbiologist is well equipped to consult with epidemiologists, infectious disease specialists and others, and to effectively contribute to the analysis and ultimate control of a nosocomial or therapeutic problem. Unfortunately, most microbiologists either fail to recognize that they can make this contribution or they believe the various kinds of biotyping are too complex and are reserved for the more sophisticated laboratories or state or federal reference laboratories. While it may be true that the more complex biotyping technics, such as serotyping of certain genera of the gram-negative enteric bacteria or phagetyping of *Staphylococcus aureus,* are best done by a reference laboratory, many biotyping methods are within the work scope of most laboratories.

Finally, biotyping offers the microbiologist and others convincing evidence that bacteria do not always lend themselves to the artificial categories we have established in our attempts to classify them. In this monograph, the practical, theoretical, and philosophical aspects of biotyping will be discussed, with the hope that more laboratories will recognize the relative ease with which certain biotyping technics can be performed and the value of biotyping in the detection and control of infectious diseases in the hospital and community.

Chapter 2

THE MEANING AND CLINICAL SIGNIFICANCE OF BIOTYPES

James D. MacLowry

THE concept of biotype in clinical microbiology is somewhat elusive and connotes many things to different investigators. At best biotyping suggests the multifaceted characteristics of groups of organisms and at worst is so general and vague in its meaning as to be a completely useless concept. The use of this term has led to often unfilfilled hope for the routine critical identification of bacteria at a subspecies or strain level, as well as to serious disputation between otherwise very rational investigators.

In the simplest sense, this concept is concerned with the genetic composition of bacteria and the hope that in some way the molecular structure of the DNA can be discerned, and organisms of like structure can be placed in the same group and critically separated from organisms of minimally or markedly dissimilar structure. Unfortunately, the state of the science and art of microbiology makes it impossible for the clinical microbiologist, and almost impossible for the research microbiologists, to consistently get an accurate picture of the genetic composition of any organism. As a consequence, it is necessary to rely on the phenotypic expression of the DNA to indirectly provide information on its structure. Since it is not possible at present to evaluate all of the physical and chemical features of a given organism, some specific characteristic or set of characteristics is studied in the hope that it will in some way be useful in providing comparisons between clinical isolates.

If one reflects for a moment on the considerable difficulty in gaining unanimity of opinion at the genus and species level, it is not at all surprising that there are even more severe definitional problems at the subspecies or strain level. From a practical standpoint, the microbiologist who has a clinical interest, whether in human or veterinary microbiology, generally is attempting to answer only a few specific questions when concerned with biotypes of a specific species. Specifically, from an epidemiologic standpoint, one wants to know whether two or more individual isolates of the same genus and species are in fact from the same point source and hence have a specific importance when considering nosocomial infections. In spite of the very considerable increases in our knowledge of the relationships between various species, there is still considerable ignorance for most hospital associated infections related to our ability to make a critical statement about the origin of the infection. A second general question which is usually asked by the clinician, has to do with whether a particular pathologic state is associated etiologically with a particular species or strain which has particular properties giving it an advantage over the host. Although a considerable amount of information has been gained in this area of study, and we are able to make a few prognostic statements about certain biotypes, pitifully little is known about the mechanisms of pathogenicity of those strains when compared to other much less virulent strains or to nonvirulent strains of the same species.

The attempts then to develop techniques for evaluating bacteria at a subspecies level are hardly frivolous academic exercises, but rather they should rightfully command the respect of the clinical microbiologist. There are, however, a number of very serious problems which need to be considered before feeling somewhat smugly that such procedures have been developed and adequately evaluated. In spite of or because of its diminutive size, the bacterium presents prodigious problems for enumeration of those characteristics which will allow us to unequivocally establish its uniqueness or similarity with another bacterium. In some ways the problem might be likened to the analysis of an aerial photograph of the Los Angeles Colosseum when USC and Notre Dame meet in their annual classic football confrontation. Given sufficient photographic resolution it may be obvious that a group of *Homo sapiens* has assembled and it might even be possible for a perceptive observer to determine the reason for the meeting. It would be much more difficult to discern each individuals feelings about the meeting and how the meeting is progressing; it would certainly not be possible to make judgements about the relationship of one individual to another. Quite different techniques would obviously be necessary for that evaluation. A bacterial population is really viewed from a similar distance, and it is not intuitively obvious which characteristic or set of characteristics is the best for study if one wants to establish similarities between populations. Further, it is conceivable that the features used to distinguish one population from another may well not be the most reasonable to distinguish either or both of them from a third population.

A number of different properties of the bacterium have been used to establish schemes for the identification of different biotypes. In this sense it may be much more correct or useful to identify organisms so grouped not as biotypes but rather more specifically to designate the type of method used for grouping, such as serotype, phagetype, etc., as has been suggested by some workers. Such a suggestion perhaps would tend to decrease the general feeling that groups developed by these different methods would necessarily include the same members.

A brief mention of some of these biotyping methods is probably germane. Some of the oldest techniques utilized a variety of biochemical test results. These tests reflected a number of different biochemical characteristics of an organism including metabolic pathways, cell wall constituents, presence of various enzymes, etc. More recently the antimicrobial susceptibility testing of bacteria has rightfully been considered an evaluation of certain biochemical characteristics of a bacterium.

A second characteristic which has been used now for over fifty years has to do with the ability of certain bacteria to undergo bacterial phage attachment and subsequent lysis. When this phage lysis phenomenon was first described there was hope that it might have therapeutic usefulness; it was only later when the technique was better defined that it became obvious it might be useful for accurately identifying various bacteria.

Considerable literature has accumulated on the ability of various bacteria to produce bacteriocins, and presently there is a certain amount of interest in the use of pyocins for the subgrouping of *Pseudomonas aeruginosa.* Colicins have been used for the grouping of certain species of Enterobacteriaceae but thus far these have had a limited usefulness.

A very considerable amount of work has been expended in the area of serology of microorganisms. The whole microorganism or many of its components have all been used as antigenic sources for a wide variety of serological schemes. The development of absorbed specific antisera to the O, H, or K antigens of the Enterobacteriaceae and to the M and T antigens of streptococci are two examples of this work. Although the actual procedures for the determination of the specific serotype are not particularly difficult, good specific antisera is in short supply; therefore these tests are performed routinely in very few laboratories. As a consequence the epidemiologic information which could be derived from this serotyping is usually not available to the vast majority of hospitals. These procedures are expensive and time-consuming, but retrospective and particularly prospective studies of bacterial populations within defined hospital groups should certainly be supported to a much greater degree than is presently occurring. Other types of serologic procedures will occasionally use whole cell extracts or will evaluate immunologically the presence of various toxic substances produced by a bacterium. All these studies are fraught with certain problems in that the antibodies are hard to purify, and therefore a certain amount of cross reactivity will occur. Also some antigens have a wide distribution making them much less useful for identification at the strain level. It is worthwhile to note at this particular juncture that different serotyping methods may yield entirely different groupings of isolates from a hospital. In the case of *Pseudomonas aeruginosa,* one set of serotyping antisera may produce only two or three different sero groups in a hospital population. Using another serotyping technique, the hospital may be found to have many more and different groups of *P. aeruginosa.* Obviously from an epidemiologic and pathogenic standpoint it might be more useful to have more groupings of organisms, but it is not necessarily obvious that this is so. One system can appear to produce too little information and the other too much, but it is a general feeling that, in attempting to identify bacteria at the strain level, more information is probably preferable to less information. There is, however, no good evidence at present as to which scheme is necessarily more useful.

Other techniques utilizing the chemical composition of the cell have been proposed and some have been attempted. The electrophoretic mobility of soluble proteins of cells can be examined as can the cell wall amino acid composition; these might yield information of some usefulness in grouping. The infrared spectrophotometer has been used to look at whole cell preparations and has some promise. Gas liquid chromotography of cell wall pyrolysis specimens may hold a great deal of promise, but these procedures are at the moment beyond the capability of most routine laboratories.

A technique which will be described later has generated a considerable amount of excitement, but whether it will turn out to be routinely useful is at present problematic. This procedure of nucleic acid hybridization is certainly one of the few which allows the investigator to look at similarities between isolates, presumably at the genetic level. The problem of this method relates to how finely one can identify the differences between isolates and how readily one can make interpretations as to whether or not two different populations of a given species are in fact different at the genetic level.

It has been our experience in attempting to make distinctions within species using large numbers of biochemical tests that we have been frustrated more often than we have had success. We have kept careful epidemiologic data of our isolates of the family Enterobacteriaceae and have compared specific problem organisms with this epidemiologic data base. Usually it has occurred that the biochemical pattern of the isolates in question have the same pattern as the most common isolates in our patient population; therefore, having many biochemical test results does not necessarily help one in feeling confident that the given isolate or isolates are from the same source. If we further add to this traditional biochemical information the quantitative antimicrobial susceptibility data, we find that a number of isolates of a given species may have the same biochemical pattern but may have quite different susceptibility patterns. Conversely, they may have the same susceptibility patterns but different biochemical patterns. We have never been certain under these circumstances which, if either, of these grouping schemes to consider as being definitive in establishing strain identity. There certainly seems to be less of a problem in interpretation if one uses a serotyping scheme, but once again we have little information on large patient populations to suggest whether or not a given serotype is commonly or uncommonly expected. The general assumption is made that if two isolates have the same serotype or phage type, they are derived from the same source and are presumably identical. Practically, this is a useful concept, but it may not be as uniquely definitive as we have thought.

There has been more success in terms of identifying point source infections with *Staphylococcus aureus* phage type patterns, but once again we may find two different isolates with the same phage pattern but with different antimicrobial susceptibility patterns. Can we with confidence state that these organisms are from the same point source or is this situation similar to the ambiguity that we find within the Enterobacteriaceae? Certainly this information would be of considerable clinical importance and interest, but we are often left with an uneasy feeling that we do not really know the answer.

In summary then, there is a very considerable necessity to attempt to critically evaluate bacterial identification at the subspecies or strain level, both from an epidemiologic and pathogenic standpoint. It should be understood that the use of the concept of biotype is an extremely general one which can refer to a variety of characteristics of an organism; also, conflicting answers from different typing procedures must be treated with respect, since the correct interpretation, other than perhaps from a purely practical standpoint, is not necessarily obvious. Until we are easily and reproducibly able to study the genetic composition of isolates, we have to accept the fact that we are only seeing imperfect reflections in the mirror of reality.

Chapter 3

BIOTYPING OF ENTEROBACTERIACEAE IN THE CLINICAL LABORATORY

Don J. Brenner

BIOTYPING is done by all bacteriologists, yet it is difficult to find an accurate definition of biotyping, biotype, or biogroup as these terms apply to bacteria. Some definitions of biotype follow:

1. A group of plants or animals with similar hereditary characteristics[18]
2. The organisms sharing a specified genotype[19]
3. A group of organisms having identical genetic but varying physical characteristics[2]

Within these definitions it is extremely doubtful that a biotype in bacteria has ever been described. This is because similarity or identity at the genetic level has not been determined.

The newer medical dictionaries have defined biotype with respect to bacteria; for example, "an infrasubspecific group of strains distinguishable from other strains of the same species on the basis of physiological characters."[15] Perhaps the most comprehensive definition is given by Cowan in *A Dictionary of Microbial Taxonomic Usage*[9]: "In higher plants a group of individuals with the same genetic constitution, but in microbiology a subdivision of a taxon; the term is used in different ways and the categories subdivided may be genera, species or serotypes. The subdivisions are made on differences in biochemical reactions, or in the ability to produce acid from various carbohydrates and alcohols. Kauffmann avoids the use of the term biotype but uses the more involved serofermentative phage type, which presumably means a biotype of a serotype and/or phage type."

Realistically, biotyping is the use of biochemical reactions to characterize, identify, classify, or differentiate bacteria. Clinical bacteriologists should all be proficient in biotyping for these applications. Biotyping is usually carried out as a means of speciation or subspeciation. A biotype can be designated arbitrarily based on differences in one reaction or twenty reactions.

It is often assumed that all unspecified reactions are shared by members of all biotypes within a given species. This is certainly not always true, and probably this is very rarely the case.

The biotyping concept is a powerful epidemiological weapon, both when used alone and when used in conjunction with serology. There are often cases where biotyping is more definitive than serotyping. This is true in several taxa including *Salmonella, Shigella,* and *Yersinia.* Moreover, there are many organisms for which there are not readily available antisera or available serotyping schemes. These include species of *Enterobacter, Citrobacter, Proteus, Providencia* and *Yersinia.*

The value of biotyping is obvious when rare types are encountered. For instance, less than 0.5 percent[10] of *E. coli* strains can utilize sodium citrate. The findings of several citrate-positive *E. coli* strains should immediately suggest the possibility of an outbreak. The value of biotyping commonly seen types is less obvious but no less important.

The purposes of this presentation are to summarize some of the information on biotypes within *Enterobacteriaceae* and to illustrate the relevance and importance of biotyping in the clinical laboratory.

Classification

Biogrouping or biotyping (these terms will be used interchangeably) is often used to identify and characterize biochemically atypical groups thought to belong to a given species. A case in point is the lactose-negative, nonmotile, anaerogenic strains of *E. coli,* once called the Alkalescens-Dispar group. These strains were shown to be a biotype of *E. coli* by biochemical[10] and then by genetic criteria.[5] *Citrobacter freundii* contains indol-positive, H_2S-negative; H_2S-negative, decarboxylase-negative biotypes, and several other biotypes.

Identification of these strains as belonging to *C. freundii* is not difficult if a sufficient number of biochemical reactions are carried out.

There are several species in which some of the presently recognized biogroups are substantially genetically different. In all probability these species will eventually be separated into two or more species on the basis of both biochemical and genetic data. The best examples of this category are *Yersinia enterocolitica* and *Enterobacter agglomerans*. There are several biotyping schemes for *Y. enterocolitica* in which typical strains are differentiated based on indol, xylose, esculin, salicin, lactose and lecithinase (Table 3-I). A biotype, negative in all of these reactions as well as several others, is further characterized by being sucrose-negative.[6,13] Two additional biotypes have been identified, both of which are rhamnose-positive.[1,3,6,8,12] One of these (Table 3-I) is also positive in tests for raffinose, melibiose and alpha-methyl-glucoside.[1,3,6,8]

Table 3-I. BIOTYPING SCHEMES FOR *Y. ENTEROCOLITICA*

Test*	Biotype				
	1	2	3	4	5
Lecithinase	+	−	−	−	−
Indol	+	+	−	−	−
Lactose	+	+	+	−	−
Xylose	+	+	+	−	−
Nitrate, Trehalose, Ornithine Decarboxylase	+	+	+	+	−

*Adapted from Wauter's[17]

Test†	Biotype				
	1	2	3	4	5
Salicin	+	−	−	−	−
Esculin	+	−	−	−	−
Indole	+	+	−	−	−
Lactose	+	+	+	−	−
Xylose	+	+	+	−	−

†Adapted from Niléhn and Sjostrom.[13]

	Biotype						
Test‡	1	1a	2	3	4a	4b	4c
Indole	−	−	+	+	+	−	−
Xylose	−	+	+	+	+	−	+
Esculin	−	−	−	+	+	−	−
Salicin	−	−	−	+	+	−	−
Lactose	−	−	−	−	+	−	+
Rhamnose	−	−	−	−	+	−	−
Sucrose	+	+	+	+	−	+	−
Orthine Decarboxylase	+	+	+	+	+	−	+

‡Adapted from Knapp and Thal.[12]

	Biogroup			
Test§	1 (typical)	2	3	4
Sucrose	+	+	+	−
Rhamnose	−	+	+	−
Raffinose	−	−	+	−
Alpha-Methyl Glucoside	−	−	+	−
Melibiose	−	−	+	−

§Adapted from Brenner et al.[6]

The two rhamnose-positive biotypes were separated from each other, and from both typical and sucrose-negative strains, by DNA hybridization.[6] The degree of DNA relatedness observed in these experiments indicated that each rhamnose-positive group belonged to a species separate from *Y. enterocolitica*. In this case the genetic basis for separation was discovered prior to the biochemical basis.

Enterobacter agglomerans is a taxonomic nightmare, and the prospect of identifying this organism has undoubtedly driven more than one clinical bacteriologist to drink or to early retirement. The organisms now known as *Enterobacter agglomerans* by clinical microbiologists were first isolated from plant sources more than seventy years ago. They have been classified as several species in the genus *Erwinia*. In the current *Bergey's Manual*[7] these organisms are placed in the Herbicola Group. There are three species, one of which has two varieties:

Erwinia herbicola var. *herbicola, E. herbicola* var. *ananas, E. stewartii,* and *E. uredovora.* These organisms were brought to the attention of clinical microbiologists in 1970 when they were responsible for a large outbreak of septicemia in contaminated intravenous solutions. Ewing and Fife characterized the outbreak strains and many other strains from animal and plant sources.[11] These authors grouped all strains into one species, *Enterobacter agglomerans,* and recognized eleven biogroups on the basis of gas formation, nitrate reduction, indol and Voges-Proskauer reactions (Table 3-II). These biogroups are sufficiently different in those reactions not used in biogrouping to make their identification extremely difficult.

Table 3-II. BIOGROUPS OF *E. AGGLOMERANS**

Biogroup	Test			
	Gas from glucose	Nitrate	Indol	Voges-Proskauer
1	−	+	−	+
2	−	+	−	−
3	−	−	−	−
4	−	−	+	+
5	−	+	+	−
6	−	−	−	+
7	−	+	+	+
G1	+	+	−	+
G2	+	+	−	−
G3	+	+	+	−
G4	+	+	+	+

*Adapted from Ewing and Fife[11]

DNA relatedness values reflect the biochemical heterogeneity seen in *E. agglomerans.** There are ten to twelve different DNA relatedness groups in strains of *E. agglomerans,* and these relatedness groups do not correlate well with the existing biogroups. It appears that neither the three species concept nor the one species, eleven biogroup concept correlates well with DNA

---

*Brenner, unpublished observations.

relatedness data. We are now trying to determine which biochemical tests correlate with DNA relatedness groups in order to (1) establish more meaningful biogroups, (2) clarify definitive reactions for identification of *E. agglomerans*, and (3) assess the clinical significance of the various biogroups. When this project is completed the group of organisms now included in *E. agglomerans* can be adequately speciated.

Identification

The failure to recognize biotypes often results in misidentification or in failure to isolate an organism. The following examples illustrate this point:

(1) Lysine⁻, arginine⁻, ornithine⁻ negative strains of *E. coli* can be mistaken for *E. agglomerans*.
(2) Anaerogenic, lactose-negative, nonmotile *E. coli* can be mistaken for Shigellae.
(3) Aerogenic *S. flexneri* or sucrose- and lactose-positive *S. sonnei* can be confused with *E. coli*.
(4) Lactose-positive *Salmonella* (or *Shigella*) are often missed entirely when lactose-positive colonies are not picked from enrichment media.

A plasmid is any self-replicating deoxyribonucleic acid structure that is not part of a chromosome. Many genetic entities fall into this category; among them are sex factors such as F-factor, bacteriophage, colicins, and R-factors which carry the genes for multiple drug resistance. In order to transfer genes throughout a bacterial population, plasmids must contain an infectious element, that is, a piece of DNA that can be transmitted from one cell to another. The F-factor is infectious and RTF (resistance transfer factor) is the transmissible agent responsible for transferring antibiotic resistance genes. There are a group of transferable plasmids that carry metabolic genes. These are called metabolic plasmids and are responsible for a large number of the atypical biotypes in members of *Enterobacteriaceae*. Epidemic strains of lactose- and/or sucrose-positive *Salmonella* are excellent examples of biotypes that are due to the presence of metabolic plasmids.

Metabolic plasmids are a mixed blessing from a diagnostic standpoint. Once identified these biotypes are very strong epidemiological markers (see below), however, the presence of the atypical gene(s) often make identification difficult. A few plasmid mediated biotypes are given in Table 3-III. Many of these plasmid mediated reactions were thought to be almost inviolate. For instance H_2S production, urea degradation, and KCN production immediately removed *E. coli* from consideration. We now know that no single reaction can a priori be considered inviolate. Identification must be based on the overall biochemical picture presented by a strain. To paraphrase Dr. W. H. Ewing, "once studied, atypical strains become typical *something.*" Sometimes they become new species as with *Citrobacter diversus,* but more often they become perfectly typical biotypes of a recognized species.

Table 3-III. SOME KNOWN OR PRESUMED METABOLIC PLASMIDS IN *ENTEROBACTERIACEAE*

Species	Plasmid Mediated Reaction
E. coli	H_2S
E. coli	Urea
E. coli	KCN
E. coli	Citrate
Typhimurium Salmonella	Sucrose
S. typhimurium	Lactose
Proteus species	Lactose
Providencia species	Urea
Providencia species	Rhamnose
Shigellae	H_2S
Serratia marcescens	Raffinose

Clinical Significance of Specific Biotypes

Marked differences in pathogenicity exist among different biogroups within a species. The biotypes of *Y. enterocolitica* that are rhamnose-negative are usually isolated from cases of gastroenteritis in humans and from sick or dead animals. The host range of the rhamnose-positive biotypes of *Y. enteroco-*

litica for animals and man is similar, but these strains are found much more often in water and foods than are their rhamnose-negative relatives. Moreover, the rhamnose-positive strains have only rarely been implicated in gastroenteritis and are usually isolated from healthy animals.[1]

In most species the clinical significance of various biotypes has not been well studied. A good example is that of *Klebsiella pneumoniae*. This organism causes about 2 percent of pneumonia cases and also can cause gastrointestinal and urinary tract infection as well as septicemia. *K. pneumoniae* is also a significant cause of nosocomial infections. The organism is quite ubiquitous in the environment, especially in vegetables, wood, and water[14] The clinical significance of these strains is not fully understood. We know that garden vegetables are often contaminated and eaten raw (lettuce, green onion, radish), apparently without ill effects. On the other hand the sale of redwood water storage tanks contaminated with *Klebsiella* has been banned (R. Seidler, personal communication). Serotyping does not resolve the problem as the same serotypes are found in environmental as well as clinical sources. Careful biotyping may help to resolve these problems.

Epidemiology

There are certain organisms where biotyping is known to be more sensitive than serotyping and both methods are routinely used to determine a bioserotype. There are 6 biotypes of *S. flexneri* 6 that are distinguished by their ability to form acid or acid and gas in the catabolism of glucose, mannitol, and dulcitol (Table 3-IV). In *Salmonella* there are many examples of serotypes that contain two or more biotypes (Table 3-V). A recent study on serotyping and biotyping in *E. coli*[16] showed that more than 20 percent of the serotypes could be subdivided into biotypes using only nineteen biochemical tests. Furthermore, approximately 15 percent of the strains were rough and therefore could not be serotyped. These investigators[16] list the following advantages of biotyping over serotyping:

1. All strains are typeable.
2. Biotyping is less laborious.

Table 3-IV. BIOTYPES OF S. FLEXNERI 6

Name	Glucose	Mannitol	Dulcitol
Boyd 88	A	A	(A) or–
Manchester	AG	AG	(AG)
(variant)	AG	AG	–
Newcastle	AG	–	(AG)
Hertfordshire	A	–	–
Sussex	A	–	(A)

A = acid formation within 2 days ( ) = delayed reaction
AG = acid and gas formation within 2 days – = negative reaction

Table 3-V. SOME *SALMONELLA* SEROTYPES THAT CONTAIN MORE THAN ONE BIOGROUP

Salmonella Type	Antigenic Formula
S. enteritidis	
ser Paratyphi-B	1,4,5,12:b:1,2
bioser Java	1,4,5,12:b:1,2
ser Neumuenster	1,4,12,27:k:1,6
ser unnamed	1,4,12,27:k:1,6
S. cholerae-suis	6,7:c:1,5
bioser Kunzendorf	6,7:c:1,5
S. enteritidis	
bioser Paratyphi-C	6,7:c:1,5
bioser Decatur	6,7:c:1,5
bioser Typhisuis	6,7:c:1,5
ser Tennessee	$6,7:z_{29}:-$
ser unnamed	$6,7:z_{29}:-$
ser Miami	1,9,12:a:1,5
ser Sendai	1,9,12:a:1,5
ser Fresno	$9,46:z_{38}:-$
ser Unnamed	$9,46:z_{38}:-$

3. Biotyping can be performed in virtually every laboratory.

Most laboratories do not have an extensive serotyping capability, and are, therefore, limited to biotyping if they are to do any primary epidemiological work. There are no good guidelines available for judging how far to go in biotyping or for choosing which tests yield definitive information for specific species. This lack of specific guidelines is good in the sense that different laboratories have different capabilities, restraints, and needs with respect to biotyping.

The simplest way to start a biotyping program is to merely inspect and analyze the data generated from a routine battery of biochemical tests. Obviously the degree of sensitivity will vary with the number and the kind of tests employed. A laboratory that routinely uses twenty tests will have a much better data base than a laboratory that uses six or eight tests. Several of the kit systems identify organisms by a profile number. These profiles are biotypes and may be profitably used to assess the incidence of various biotypes of a given species. Reactions that vary within a species are useful in biotyping, especially when several of these are compared. Reactions that rarely vary should also be included, because an organism may often be biotyped almost solely on the basis of an atypical reaction in one of these tests.

Biotypes for a hypothetical species of *Enterobacteriaceae* are given in Table 3-VI. Three of the thirteen reactions shown, H_2S, urea, and KCN are of no value in biotyping, because all strains give the same reaction (H_2S-positive, urea-negative, and KCN-negative). The remaining reactions divide this species into ten biotypes. Some of these will rarely be encountered (biotypes C, D, G, H, I, J) because they exhibit reactions seen in only 1 to 4 percent of all strains. The remaining biotypes (A, B, E, F) will make up the large majority of isolates. Multiple isolates of a rare biotype from different patients are usually strong presumptive evidence for an outbreak. In such cases the source of isolates should be checked and attempts made to determine whether a common source of infection exists.

Table 3-VI. BIOTYPES OF A HYPOTHETICAL SPECIES
OF *ENTEROBACTERIACEAE*

Reaction	% Positive for all strains	Biotype									
		A	B	C	D	E	F	G	H	I	J
Indol	99	+	+	+	+	+	+	+	+	+	−
Methyl Red	1	−	−	−	−	−	−	−	−	+	−
V-P	93	+	+	+	+	+	+	+	−	+	+
Citrate	98	+	+	+	+	+	+	−	+	+	+
H_2S	100	+	+	+	+	+	+	+	+	+	+
Urea	0	−	−	−	−	−	−	−	−	−	−
KCN	0	−	−	−	−	−	−	−	−	−	−
Lysine	45	+	−	+	−	+	−	+	+	−	+
Arginine	57	+	+	−	+	−	+	−	+	−	+
Ornithine	62	+	+	+	+	−	+	−	+	+	+
Lactose	96	+	+	+	−	+	+	+	+	+	+
Sucrose	2	−	−	+	−	−	−	−	−	−	−
Adonitol	80	+	−	+	+	+	+	+	−	+	+

The epidemiological significance of commonly encountered biotypes is perhaps less obvious, but nonetheless of great value, especially with respect to nosocomial outbreaks. Let us consider the four common biotypes of our hypothetical species in a bit more detail. A hospital laboratory has been compiling biochemical data together with source of isolation and frequency for our hypothetical species (Table 3-VII). There are several ways in which these data can be used. The appearance of any biotype from an unusual source is suspicious. For example, this organism is very rarely isolated from CSF or sputum. If five cases appeared during the course of a month, a thorough investigation should be made. During an average month there are twenty-three stool isolates. If thirty-five isolates are made, biotyping should be checked to see whether each type has been isolated with increased frequency or whether the entire increase is due to one type. Ten isolates from urine are encountered in an average month. If fourteen isolates are found, biotyping may indicate that ten of these are biotype A, a four-fold increase.

Table 3-VII. SOURCE AND INCIDENCE OF COMMON BIOTYPES OF A HYPOTHETICAL SPECIES OF *ENTEROBACTERIACEAE*

	Biotype*			
	A	B	E	F
Blood	0.6	0.4	0.7	0.6
Urine	2.5	3	2	2.5
Stool	6	4	8	5
Sputum	0	0	0	0.1
Wound	1	0.2	0.5	0.8
CSF	0	0	0.1	0

*Average number of independent isolates per month over the period of 2 years.

In these situations biotyping is of great help in detecting nosocomial outbreaks at an early stage. It is always possible to seek confirmation by serotyping, if available, or by additional biochemical reactions. The purpose of this paper is not to present biotyping as a cure-all, because it is not. Biotyping is a flexible tool for identification as well as for epidemiological surveillance. As such it can be used in conjunction with or independently from other epidemiological aids. Biotyping is often as sensitive or more sensitive than serotyping and has the advantage that it can be used by every microbiology laboratory without additional reagents or equipment.

REFERENCES

1. Alonso, J. M., J. Bejot, H. Bercovier and H. H. Mollaret. Sur un groupe de Souches de *Yersinia enterocolitica* fermentant le rhamnose. Intérêt diagnostique et particularités écologiques. *Méd et Maladies Infect,* 5:490-492, 1975.
2. *The American Heritage Dictionary of the English Language.* New York, Houghton Mifflin Co., 1970.
3. Bottone, E. J., B. Chester, M. S. Malowany and J. Allerhand. Unusual *Yersinia enterocolitica* isolates not associated with mesenteric lymphadenitis. *Appl Microbiol,* 27:858-861, 1974.
4. Brenner, D. J. *Shigella flexneri* 6 biotypes: a review. Atlanta, Center for Disease Control Publication, 1975.
5. Brenner, D. J., G. R. Fanning, F. J. Skerman and S. Falkow. Polynucleotide sequence divergence among strains of *Escherichia coli* and closely related organisms. *J. Bacteriol,* 109:953-965, 1972.
6. Brenner, D. J., A. G. Steigerwalt, D. P. Falcão, R. E. Weaver, and G. R. Fanning. Characterization of *Yersinia enterocolitica* and *Y. pseudotuberculosis* by DNA hybridization and by biochemical reactions. *Int J Syst Bacteriol,* 26:180-194, 1976.
7. Buchanan, R. E. and N. E. Gibbons. *Bergey's Manual of Determinative Bacteriology,* 8th Ed. Baltimore Williams & Wilkins, 1974.
8. Chester, B. and G. Stotzky. Temperature-dependent cultural and biochemical characteristics of rhamnose-positive *Yersinia enterocolitica.* *J Clin Microbiol,* 3:119-127, 1976.
9. Cowan, S. T. *A Dictionary of Microbial Taxonomic Usage.* Oliver and Boyd Ltd., 1968.
10. Edwards, P. R. and W. H. Ewing. *Identification of Enterobacteriaceae.* Dallas, Burgess Publ. Co., 1972.
11. Ewing, W. H. and M. A. Fife. *Enterobacter agglomerans.* Atlanta, Publication of the Center for Disease Control, 1971.
12. Knapp, W. and E. Thal. Differentiation of *Yersinia enterocolitica* by biochemical reactions. *Contrib Microbiol Immunol,* 2:10-16, 1973.
13. Niléhn, B. and B. Sjostrom. Studies on *Yersinia enterocolitica. Acta Pathol Microbiol Scand [Suppl]* 206:1-48, 1969.
14. Seidler, R. J., M. D. Knittel and C. Brown. Potential pathogens in the environment: cultural reactions and nucleic acid studies on *Klebsiella pneumoniae* from clinical and environmental sources. *Appl Microbiol,* 29:819-825, 1975.
15. *Stedman's Medical Dictionary,* 22nd Ed. Baltimore, Williams & Wilkins, 1972.
16. van der Waaij, D., T. M. Speltie, P. A. M. Guinee and C. Agterberg. Serotyping and biotyping of 160 *Escherichia coli* strains: comparative study. *J Clin Microbiol,* 1:237-238, 1976.

17. Wauters, G. Contribútion à l'étude de *Yersinia enterocolitica*. Thesis, 165 pp. (Vander, Louvain), 1970.
18. Webster's New World Dictionary of the American Language. Mountain View, Calif., World Publ. Co., 1966.
19. Webster's Seventh New Collegiate Dictionary. Springfield, Mass., G. and C. Merriam Co., 1965.

Chapter 4

BIOTYPING OF NONFERMENTATIVE, GRAM-NEGATIVE BACTERIA

J. J. Farmer III

BIOTYPING of the nonfermentors is a difficult subject to cover because of the incomplete and often confusing taxonomy in this group of bacteria. There is considerable variation in the biochemical reactions of the species that comprise this group.[14] In this chapter I will point out some of the difficulties with the nonfermentors as evidenced by my experience with the fluorescent pseudomonads.[8,12]

The classification of the fluorescent pseudomonads in the 8th Edition of *Bergey's Manual of Determinative Bacteriology* is shown in Table 4-I.[5] There can be many biochemical varieties or biovars among the different species (Table 4-II).

Table 4-I. SPECIES AND BIOVARS IN THE FLUORESCENT PSEUDOMONADS: THEIR CLINICAL RELEVENCE

Species or Biovar	Frequency in Clinical Specimens and Human Disease
P. aeruginosa	Very important pathogen
P. putida	
P. fluorescens	
Biovar I	Occasionally found, not often the cause of disease
II	
III	
IV	
Miscellaneous strains	
P. chloraphis	Rarely found
P. aureofaciens	
P. stringae	
P. cichorii	Extremely rare, if ever

Table 4-II. VARIATION IN THE BIOCHEMICAL PROPERTIES OF 3 SPECIES*

Biochemical Test	Species: P. aeruginosa	P. fluorescens	P. putida
Motility	93†	100	100
Urease	23	40	51
Pyocyanin production	58	0	0
Nitrate→ Gas	94	2	0
Arginine dihydrolase	96	98	97
Oxidation of: Maltose	0	70	35
Lactose	0	26	28
D-Mannitol	81	94	19
L-Rhamnose	6	84	63
Sucrose	12	64	13

*Data is from Lennette et al.[14]
†The numbers give the % positive within 2 d.

Although *Pseudomonas aeruginosa* is a species with variations in many of its properties, in its biochemical reactions, it is a moderately tight species; however the table shows that there are a number of different biovars possible — non-motile, urea+, surcrose+, plus various combinations of the variable characters.

In Figure 4-1 *P. aeruginosa* is diagramatically shown as a species with several distinct biovars. The peaks on the curve represent distinct groups in the species. The first peak represents the mucoid biovar of *P. aeruginosa* which is usually obtained from children with cystic fibrosis or other chronic respiratory infections.[7] The second peak represents biovars which are now identified as "apyocyanogenic strains of *P. aeruginosa*"[14] This group may represent unusual biovars or mutants of *P. aeruginosa* or they may represent one or more new species which have not been described. The third peak represents bioserovar 11, which has caused all of the known outbreaks of whirlpool-associated skin rash.[6] A fourth peak which is probably not part of the species *P. aeruginosa* represents a group of pseudomonads which grow at 41C and produce fluorescent pigment.[11] The taxonomic position of this group is unclear. The most impressive aspect of variations in *P. aeruginosa* is how quickly the strains can change. During a nursery outbreak,[3] a single strain infected nine different infants and contaminated twelve environmental reservoirs. The serotype and pyocin production pattern of the epidemic strain remained constant during the outbreak, but there were changes in the following characteristics: sulfadiazine sensitivity, shade of green pigment produced, fluorescein production, autoplaque production, hemolysis, bacteriophage production, and pyocin sensitivity. All these changes took place within a few weeks and illustrate the phenomena — "refluxing biovars." This is my definition of biovars of the same strain that can rapidly change as a function of human, environmental, or other selective pressures.

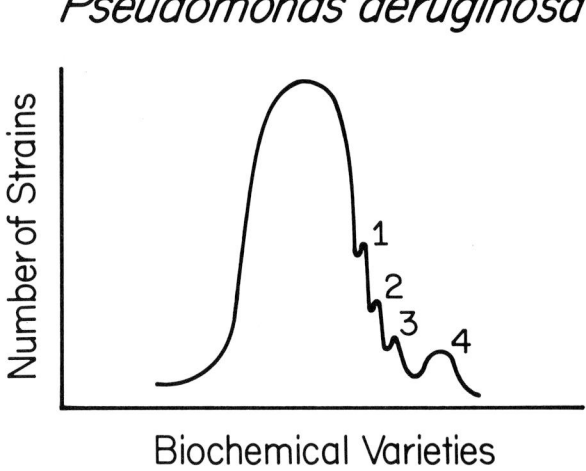

Figure 4-1. Graphic representation of the distinct groups in *Pseudomonas aeruginosa*.

We can equate the biovars of *P. aeruginosa* to the races of man, or, in the context of this book, to the biovars of man. Both *P. aeruginosa* and *Homo sapiens* form well-defined species, but each species has considerable phenotypic variation which is expressed as different biovars in *P. aeruginosa* and as different races or gene pools in man. Many bacterial species can also be considered in this way; however, it is often difficult to tell the difference between a "variable species," with many different biovars, and "tight species," which are very closely related to each other. This difficulty is apparent in *Pseudomonas fluorescens*.

P. fluorescens also gives variable results on tests normally used in clinical microbiology (Table 4-II). However, on the basis of the more extensive characterization used by Stanier, Palleroni, and Doudoroff,[16] *P. fluorescens* can be broken into at least five biovars. It is difficult to distinguish a biovar of *P. fluorescens* from what may be a true species. Two different ways to interpret the biochemical variation in this "species" are shown in Figure 4-2. The diagram on the left shows *P. fluorescens* as a single species with considerable variations in its biochemical properties. It contains biovars I-IV and many additional biovars which are grouped together as "miscellaneous strains." This is the current classification given in the 8th Edition of *Bergey's Manual of Determinative Bacteriology*. An alternative classification is given in Figure 4-2, in which some of the named and unnamed biovars are elevated to the level of species. DNA hybridization studies are beginning to clarify the taxonomy of the "*P. fluorescens* complex." A similar classification problem has been seen with the species "*Enterobacter agglomerans*" in the family Enterobacteriaceae. What appeared to be a single species is made up of over a dozen hybridization groups which should eventually be elevated to the status of species.[4]

The differentiation of "biovars" from "true species" is going to be a problem for some time to come. The biochemical data from the nonfermentors can be interpreted in both ways: The "species lumpers" would say that there are many biovars in species with biochemical variations, but the "species splitters"

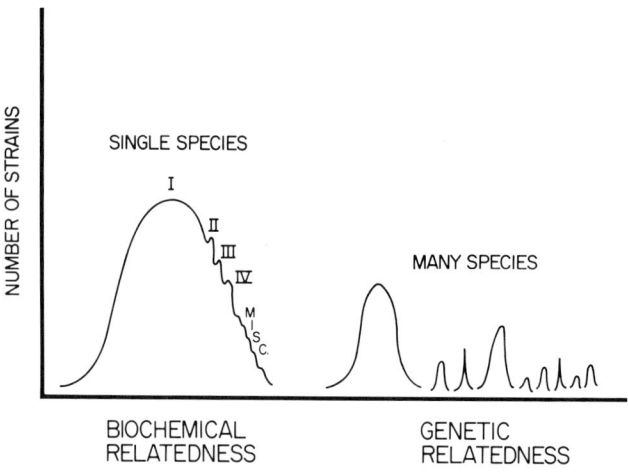

Figure 4-2. Two different interpretations of the biochemical variation in *Pseudomonas fluorescens*; left: *P. fluorescens* as a single species with many biovars; right: the biovars classified as distinct species.

would say that the biovars should be made species. DNA hybridization in the best characterized of all bacterial groups — the Enterobacteriaceae — has shown that both points of view will probably prove to be correct; however, it is not possible to predict the genetic groupings in advance.[4] It may turn out that the correct taxonomy will give insight into the importance of the biovars or species in human disease and public health. There is much to learned about *Pseudomonas, Moraxella, Acinetobacter, Alcaligenes, Achromobacter, Flavobacterium,* and the unnamed groups of nonfermentors.[1,2,9,10,11,13,14,16]

Based on data from the fluorescent pseudomonads, three hypotheses can be made about biovars in other groups of nonfermentative gram-negative rods.

Hypothesis 1. The species are variable in their biochemical properties and thus are made up of many different biovars; however, they are still true species. This would be analogous to the variations in man or like *Escherichia coli* in bacteria.

Hypothesis 2. What we now recognize as a single variable species is really made up of several species, but this fact is obscured because our existing classification is inadequate. Such species would be analogous to *Enterobacter agglomerans* in the Enterobacteriaceae, which is not a true species but really over a dozen species.[4]

Hypothesis 3. (I am really being a heretic here.) Some species of nonfermentors are really species of Enterobacteriaceae which have lost the ability to transport carbohydrates. These "transportless mutants" are well known in *Escherichia coli*, but we have no idea at what rate they occur naturally. An *E. coli*, *Proteus*, or *Klebsiella* which cannot transport sugars cannot ferment them during catabolism and would become a "nonfermentative, gram-negative rod."

My guess is that all three of these hypotheses are correct. Our task is to determine which hypothesis is correct for each group or species. In biotyping and taxonomy of the nonfermentors, we may be like the butterfly taxonomist working in the tropical rain forest who sees a great deal of phenotypic variation but must differentiate true species from their Müllerian mimics.[15]

REFERENCES

1. Baumann, P., M. Doudoroff, and R. Y. Stanier. Study of the Moraxella Group. I. Genus *Moraxella* and the *Neisseria catarrhalis* Group. *J Bacteriol*, 95:58-73, 1968.
2. Baumann, P., M. Doudoroff, and R. Y. Stanier. A study of the *Moraxella* Group. II. Oxidative-negative species (Genus *Acinetobacter*). *J Bacteriol*, 95:1520-1541, 1968.
3. Bobo, R. A., E. J. Newton, L. F. Jones, L. H. Farmer, and J. J. Farmer, III. Nursery outbreak of *Pseudomonas aeruginosa*: Epidemiological conclusions from five different typing methods. *Appl Microbiol*, 25:414-420, 1973.
4. Brenner, D. J. Clinical and taxonomic applications of DNA hybridization. *Public Health Lab*, 34:48-55, 1976.
5. Buchanan, R. E. and N. E. Gibbons (eds.). *Bergey's Manual of Determinative Bacteriology*, 8th ed. Baltimore, Williams & Wilkins Co., 1974.
6. Center for Disease Control. Skin rash associated with pool exposure — Minnesota. *Morbidity and Mortality Report*, 24:166-171, 1975.
7. Doggett, R. G., G. M. Harrison, and R. E. Carter, Jr. Mucoid *Pseudomonas aeruginosa* in patients with chronic illness. *Lancet*, 1:236-237, 1971.

8. Farmer, J. J., III. *Pseudomonas* in the hospital. *Hosp Pract, 11*:63-70, 1976.
9. Gilardi, G. L. Practical schema for the identifications of non-fermentative gram negative bacteria encountered in medical bacteriology. *Am J Med Tech, 38*:65-71, 1972.
10. Gilardi, G. L. Diagnostic criteria for differentiation of *Pseudomonas* pathogenic for man. *Appl Microbiol, 16*:1497-1502, 1968.
11. Hoadley, A. W., L. Ajello, and N. Masterson. Preliminary studies of fluorescent pseudomonads capable of growth at 41 C in swimming pool waters. *Appl Microbiol, 29*:527-531, 1975.
12. Jones, L. F., E. T. Thomas, J. D. Stinnett, G. L. Gilardi and J. J. Farmer, III. Pyocin sensitivity of *Pseudomonas* species. *Appl Microbiol, 27*:288-289, 1974.
13. Juni, E. Interspecies transformation of *Acinetobacter*: genetic evidence for a ubiquitous genus. *J Bacteriol, 112*:917-931, 1972.
14. Lennette, E. H., E. H. Spaulding and J. P. Truant (Eds.). *Manual of Clinical Microbiology.* Washington, D.C., American Society for Microbiology, 1974.
15. Papageorgis, C. Mimicry in neotropical butterflies. *Am Scient, 63*:522-532, 1975.
16. Stanier, R. Y., N. J. Palleroni, and M. Doudoroff. The aerobic pseudomonads: a taxonomic study. *J Gen Microbiol, 43*:159-271, 1966.

Chapter 5

STREPTOCOCCAL GROUPS OF CLINICAL SIGNIFICANCE

Hazel W. Wilkinson

Introduction

IN 1933, Dr. Rebecca Lancefield described a system for serologically classifying streptococci.[17] A group-specific, carbohydrate antigen ("C substance") was extracted from whole streptococcal cells with hot hydrochloric acid and was allowed to react in precipitin tests with antisera raised in rabbits with formalinized, whole-cell vaccines. Five different serological groups were found among the streptococci commonly isolated at that time from human, animal, and dairy sources. An additional two groups were described in 1935,[21] a total of eleven by 1940,[19] and an additional nine streptococcal groups based on the Lancefield technique have been proposed by other investigators since that time.

The Center for Disease Control (CDC) routinely uses grouping antisera for groups A, B, C, D, F, and G to classify the beta-hemolytic streptococci, and antisera for groups B and D to identify serologically alpha-hemolytic or nonhemolytic strains presumptively identified as group B or D.[8] Other alpha-hemolytic or nonhemolytic streptococci are classified by means of physiological tests, since experience gained with many thousands of these organisms has shown that considerable error occurs when they are classified serologically.[7] These errors presumably occur because of the numerous cross-reactions among their constituent type antigens. Furthermore, beta-hemolytic streptococci belonging to serological group E and beyond group G are rarely isolated from human infections and, indeed, controversy exists over their proper classification.

Group A Streptococci

Group A streptococci constituted 80 percent of all streptococci grouped at CDC in fiscal year 1969 (Fig. 5-1). These organisms cause a wide variety of acute illnesses (Fig. 5-2) that range in incidence levels from extremely high, such as the 474,000 cases of streptococcal pharyngitis reported to CDC on an optional basis by forty-six states in 1973,[24] to quite low, such as the fulminant sepsis-gangrene cases that Quintiliani and Engh reported.[26] But the uniqueness of group A may be attributed to the fact that some throat and skin infections can lead to nonsuppurative sequelae several weeks after the primary infection. Acute rheumatic fever (ARF) and acute glomerulonephritis (AGN) sometimes occur after an untreated group-A pharyngitis, a primary reason for the American Heart Association's recommendation that patients with a group-A-streptococcal pharyngitis receive antibiotic treatment.[2] Although impetigo caused by group A can also lead to AGN, according to available information it does not lead to ARF.

The sequence of events leading to ARF or to AGN is not well understood. Each disease is probably the result of an autoim-

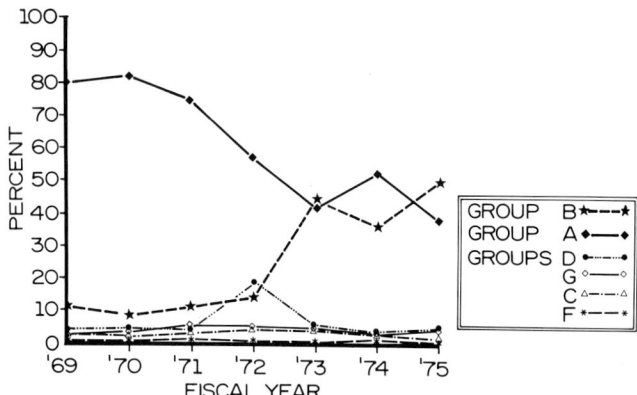

Figure 5-1. Relative percent distribution of streptococcal serological groups sent to the Center for Disease Control from fiscal year 1969 through fiscal year 1975.

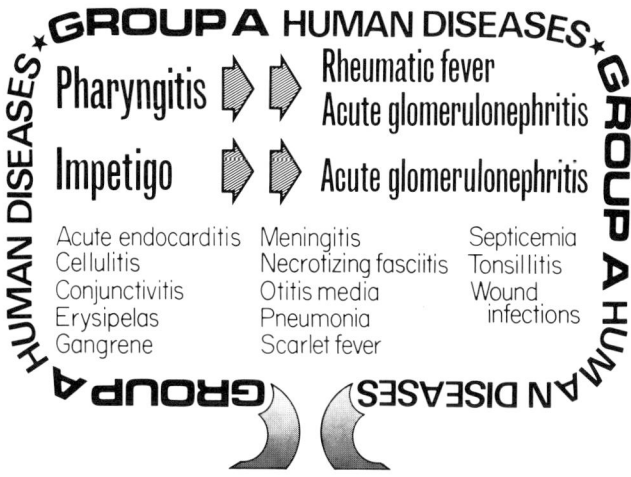

Figure 5-2. Examples of the diversity of diseases caused by group-A streptococci.

munity which may be due to cross-reactive streptococcal and host tissue antigens. Group A streptococci produce several extracellular products that may enhance the organisms' pathogenicity or provide the initial insult to host tissue and, in addition, often produce a surface-protein antigen, the M protein,* which prevents the streptococci from being phagocytized. Antibodies against M protein are protective in mice against challenge by the homologous-type strain, and it is probable that human immunity to streptococcal infection is also type-specific.[15]

In addition to being biologically significant, the M protein antigens are used as epidemiological markers. There are over sixty recognized M types, which are identified in the capillary precipitin test with the same HCl extracts used for grouping the streptococci. Unfortunately, less than half of the group-A strains isolated from human sources are M-typeable with available antisera, and for this reason another typing system — T

---

*So-called because it was originally thought that mucoidal or matt colonies were the result of M protein production. Dr. Armine Wilson showed, however, that colonial type was related to hyaluronic acid rather than to M-protein production.[36]

agglutination — comes in handy because most group A strains are T-typeable.* A characteristic combination of M- and T-typing reactions is a very useful epidemiological marker (Table 5-1). Certain T types or T-typing patterns may be associated with from one to a dozen or so M types. Furthermore, some M types characteristically cause porcine serum to turn opaque and, therefore, are "serum opacity reaction positive," whereas others are "serum opacity reaction negative."[23] Since the reaction may be inhibited by M-type-specific antiserum, a third serological test, called the serum-opacity reaction (SOR), may be used to "biotype" group A-streptococci. A recently described bacteriophage-typing system may have useful applications in further differentiating strains within certain M types.[29] Therefore, the entries in Table 5-I will no doubt be more numerous as new systems are developed and new M types are discovered.

Eight of seventy-one consecutive M-type numbers are not listed in Table 5-I. They were dropped from the classification scheme after M types 7, 20, and 21 were found to belong to serological group C; M type 16, to group G; and after M types 10, 35, 45, and 64 were found to be the same as M types 12, 49, 24, and 52, respectively. This leaves a total of sixty-three M protein markers and seventeen T agglutination patterns, the various combinations of which provide over 100 biotypic markers for group-A streptococci. Not bad for one serological group!

However, a large number of specific antisera are required to type group-A streptococci, the antisera are expensive to produce, the tests are time-consuming to perform, and the technicians must be specially trained to use and interpret the tests. Therefore, the tests should be used only in epidemiological studies, treatment evaluations, and special research projects designed to give the most complete information regarding host-organism interactions and, ultimately, patient management.

---

*T antigens, unlike M antigens, apparently have no effect on phagocytosis. T antibodies are not mouse protective. Similarly, a third group of protein antigens, the R proteins, are neither protective nor are they very useful epidemiological markers, since they are not type or group specific.

Table 5-I. USEFUL COMBINATIONS OF M AND T ANTIGENS AND SERUM
OPACITY FACTORS AS BIOTYPIC MARKERS OF
GROUP-A STREPTOCOCCI.*

T agglutination pattern	Associated M types† SOR‡ positive	SOR‡ negative
1	68	1
3/13/B3264	13	3,32,33,37,39,41, 42,43,52,53,69
3/13/B3264/28/9		34,56,67,69,71
2	2	
4	4,22,28,48,60,63	24,26,29
6		6
28	28,48	29,56,70
11	11	
12	22,62,66	12
5/27/44	27,44	5,70
14	49	14,51
8/25/Imp 19	2,8,25,58,59	31,55,57,65
22	22	
23		15,17,19,23,30,47,54
9	9	
18		18,36
9/11	9,11,61	
None		38,40,46,50

*Based on Maxted[23] and Facklam [9]
†M types 61 through 71 are proposed new M types with provisional designations.
‡Serum opacity reaction.

A goal of several such research projects has been to purify and characterize the M-protein antigens [5,6,10,11,14,25,27] in an effort to provide some insight into host immunity and mechanisms leading to the sequelae of a group-A streptococcal infection. The separation of opsonic from precipitating activity in several of these studies has suggested that a search for protective antibody in human sera must utilize tests that do not measure precipitating antibody alone. A recently developed enzyme-linked immunosorbent assay (ELISA) for M antibodies [28] meets this criterion, and in addition, has the advantages of being extremely sensitive and of differentiating immunoglobulin classes that bind to the antigen. Furthermore, results from the ELISA and the *in vitro* bactericidal assay for M-type-12 antibodies in both rabbit and human sera were comparable.[28] Thus, the ELISA may prove to be a good tool for obtaining some insight into immunity to group-A infections. These infections are no less important to the clinical microbiologist today than in fiscal year 1969, despite the fact that group A no longer constitutes almost all the beta-hemolytic streptococci seen at CDC (Fig. 5-1).

Group B Streptococci

The fact that the proportion of Group-A streptococci sent to CDC in this decade has decreased relative to the proportion of group-B streptococci sent concurrently (Fig. 5-1) may be explained in two ways. First, there has been an apparent increase in the incidence of group-B disease in humans, and second, CDC has provided serological typing of group-B streptococci since 1967. Group A- and B-typing are used for the same reasons — epidemiology, identification of possible treatment failures, and research studies — and group B typing is carefully conserved for the same reasons that M- and T-typing are, i.e. expense, time, and availability of trained personnel.

Group B contains five serological types: Ia, Ib, Ic, II, and III.[18,20,32,35] In addition to the group-B-carbohydrate (CHO) antigen, each type contains a type-specific CHO antigen,* and some type strains also contain protein antigens† (Table 5-II). In addition, the Ia and Ib CHOs often elicit "cross-reactive" Iabc antibodies that bind antigenic determinants common to types Ia, Ib, and Ic.[22,31] The chemical composition, molecular size, and number of antigenic determinants found in the type antigens depend on the methods used to extract and purify the antigens.

Table 5-II. CARBOHYDRATE AND PROTEIN ANTIGENS AS BIOTYPIC MARKERS OF GROUP-B STREPTOCOCCI.

Group B type designation	Antigens	
	CHO	Protein
Ia	Ia	
Ib	Ib	Ibc*
Ic†	Ia	Ibc
II	II	
III‡	III	

*Formerly designated Ic antigen.[32]
†Formerly designated type Ii.[35]
‡Type II and III strains that give positive precipitin reactions with anti-Ibc serum are designated types II/Ic or III/Ic at the present time.

The group-B-type antigens are analogous to the group-A, M-type antigens in several ways. First, they are located on the cell surface and therefore may impede phagocytosis *in vivo*.

---

*Types Ia and Ic contain a serologically and chemically identical Ia-CHO antigen that is present in all strains of these two types.
†Types Ib and Ic contain serologically similar Ibc-protein antigens that are present in all strains of these two types. Approximately 40 percent of type-II strains and, very rarely, type-III strains also contain Ibc antigens.

Second, antibodies specific for any one of several antigens, CHO, or protein, in the same type strain passively protect mice against challenge by any strain containing the antigen,[22] indicating that antisera may be *multiply* protective perhaps to a greater extent than are M antibodies. Third, the group-B-type antigens may be used as epidemiological markers but, in contrast with results obtained in the M-typing system, over 99 percent of the group B strains are typeable.[31]

The results of typing over 4,000 group-B strains at CDC indicate that all five types are associated with human disease. In adults, the organisms can cause a variety of disorders (Fig. 5-3), but the fact that 83 percent of 243 blood cultures and 93 percent of 153 cerebrospinal fluid cultures were isolated from patients who were less than two years old* illustrates the seriousness of group-B neonatal disease.

Two forms of group-B neonatal disease have been described. The early- or acute-onset syndrome is characterized by symptoms of sepsis and respiratory distress within the first few hours or several days after birth.[3,12] Since this syndrome has a high

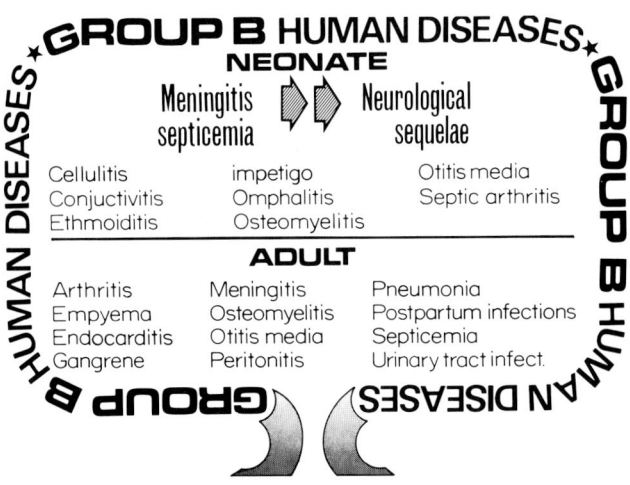

Figure 5-3. Examples of the diversity of diseases caused by group-B streptococci.

---

*Unpublished data.

mortality rate, the streptococci are often identified postmortem. Babies apparently acquire the organism congenitally, and several pertinent risk factors have been identified: maternal genital colonization, prolonged rupture of membranes prior to delivery, and low birth weight.[1,3] Group-B isolates from the blood of infants with early-onset disease, sterile CSF, and no overt clinical signs of meningitis are divided among types I, II, and III; this is as would be expected with congenitally acquired infection, since this type distribution is also seen in studies of vaginal cultures.[33] In contrast, group-B isolates from the spinal fluid of early-onset infants are predominantly type III.[3,33] This fact raises the question of whether early-onset meningitis is acquired noncongenitally, or whether congenitally acquired type III strains differ in virulence from other types.

The latter explanation seems more likely to apply to both early- and late-onset group-B disease. In late-onset group-B disease, type III predominates in both blood and CSF isolates from infants over ten days of age.[3,33] clearly, the epidemiology of late infections differs from that of early infections. One factor identified in the transmission of late-onset group-B disease is nosocomial spread.[1,30] There are undoubtedly many other possible sources of infection since vaginal colonization rates as high as 29 percent were found in studies designed for optimal recovery of the organisms.[1] Similar rates have been found in urethral cultures of sexually active males,* a fact that suggests sexual activity as another epidemiological factor in the transmission of group-B disease. Limited studies suggest that antibiotic prophylaxis may not prevent sexual transmission of the organism in all cases.[12,37]

There is also the question of whether group-B neonatal disease can be effectively controlled by antibiotic prophylaxis of colonized women or their babies.[13,30] As an alternative control method, passive protection of the fetus with maternal antibody induced by immunization with a type III CHO vaccine has been suggested.[4] However, many questions regarding the purity, immunogenicity, and safety of the antigens, as well as

---

*Dr. Stephen Swartz, personal communication.

the mechanisms of protection in humans, remain unanswered.

A radioimmunoassay (RIA) has been developed[34] to determine whether human sera contain type-specific group-B antibodies, and also whether the antibody levels in sera from healthy babies and those with a group B infection differ, as suggested recently.[4] In this test, purified type CHO antigens from all five group B types are used. As shown in Figure 5-4, each antigen is labeled with ^{125}I and is then added to the test serum. Ammonium sulfate is added to precipitate all antibody and any antigen bound to it. After removing some of the supernatant fluid, the analyst counts the remaining radioactivity and calculates the percentage of antigen bound by the antibody from the counts of ^{125}I and those of a ^{22}Na volume marker. Type-specific rabbit sera bind the homologous-type CHO when they are diluted to well over 1:1000. Sera that contain antibodies against determinants common to several types also bind accordingly in the RIA. For this reason, RIAs of human sera must be cautiously interpreted.

Conclusion

The omission of a discussion of streptococcal groups C, D, F,

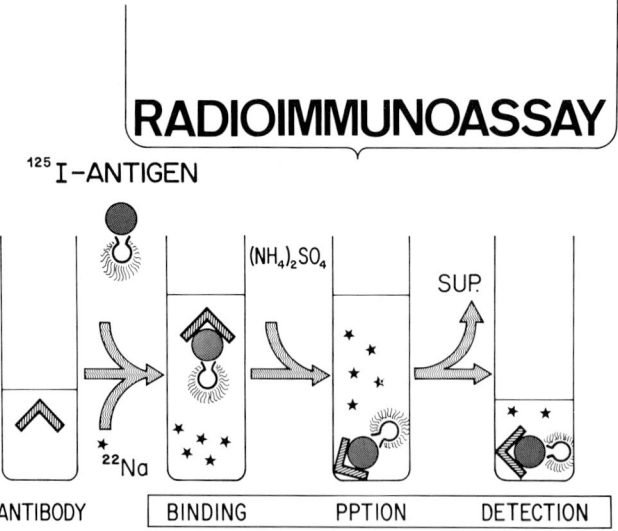

Figure 5-4. Diagram of the radioimmunoassay for measuring antibodies specific for the type carbohydrate antigens of group-B streptococci.

and G in this paper was intentional but was not meant to imply that these groups are insignificant. Enterococci, which are members of, but do not totally comprise, group D are well known to the clinical microbiologist not only because they are frequently isolated from patients with endocarditis, peritonitis, and infections of the urinary, biliary, and intestinal tracts, but also because they show greater resistance to penicillin therapy than do other streptococci. Similarly, groups C, F, and G are sometimes isolated from human pathological material, and group E as well as some of the higher lettered groups are of veterinary significance. Limited space, however, precludes a thorough review of biotypic markers of all streptococcal groups of importance to clinical microbiology, since "very few bacterial species have as much known about their chemical composition and immunochemical properties as the streptococci which have been studied by so many different investigators."[16] I therefore have discussed in some depth only the biotypic markers of groups A and B.

Several facts influenced the decision to limit this report to groups A and B.

(1) An analysis of the percentage of each group sent to CDC indicated that groups A and B comprised approximately 80 percent of the total. One reason for this distribution is that techniques for identifying biotypic markers within these two groups are available.

(2) Some of the biotypic markers of groups A and B are *biologically* significant. The M-protein-type antigens of group A and the CHO- and Ibc-protein-type antigens of group B elicit antibodies that are passively protective in mice, presumably by virtue of their opsonic activity.

(3) These biologically significant biotypic markers are identified in the laboratory with type-specific antisera. The serological typing systems in each of these two groups have given us much information regarding the epidemiology of streptococcal infections and the relationships of certain types to disease syndromes.

(4) Current research on each group includes the use of highly sensitive antibody tests — the ELISA for group-A, M-type

antibodies, and the RIA for group-B, CHO-type antibodies. Undoubtedly, other factors influence the host-parasite relationship, but a measure of specific antibodies with purified antigens in carefully controlled tests should provide information that, at least, will allow us to make rational decisions regarding future research and, at best, will give us some insight into ways to protect ourselves against these ubiquitous bacteria.

REFERENCES

1. Aber, R. C., N. Allen, J. T. Howell, H. W. Wilkinson, and R. R. Facklam. Nosocomial transmission of group B streptococci. *Pediatrics*, 58:346-353, 1976.
2. American Heart Association. *Statement on Prevention of Rheumatic Fever*, No. EM 113, 1972.
3. Baker, C. J., F. F. Barrett, R. C. Gordon, and M. D. Yow. Suppurative meningitis due to streptococci of Lancefield group B: a study of 33 infants. *J Pediat*, 82:724-729, 1973.
4. Baker, C. J., and D. L. Kasper. Correlation of maternal antibody deficiency with susceptibility to neonatal group B streptococcal infection. *New Engl J Med*, 294:753-756, 1976.
5. Besdine, R. W., and L. Pine. Preparation and description of high-molecular-weight soluble surface antigens from a group A streptococcus. *J Bacteriol*, 96:1953-1960, 1968.
6. Cunningham, M. W., and E. H. Beachey. Peptic digestion of streptococcal M protein. I. Effect of digestion at suboptimal pH upon the biological and immunochemical properties of purified M protein extracts. *Infect. Immun*, 9:244-248, 1974.
7. Facklam, R. R. Physiological differentiation of viridans streptococci. *J Clin Microbiol*, 5:184-201, 1977.
8. Facklam, R. R., J. F. Padula, L. G. Thacker, E. C. Wortham, and B. J. Sconyers. Presumptive identification of group A, B, and D streptococci. *Appl Microbiol*, 27:107-113, 1974.
9. Facklam, R. R., J. F. Padula, and E. C. Wortham. Group A streptococcal M-typing by a screening method based on T-typing results. *Abstr An Meet Am Soc Microbiol E58*, 23-28, April 1972.
10. Fischetti, V. A., E. C. Gotschlich, G. Siviglia, and J. B. Zabriskie. Streptococcal M protein extracted by nonionic detergent. I. Properties of the antiphagocytic and type-specific molecules. *J Exper Med*, 144:32-53, 1976.
11. Fox, E. N. M proteins of group A streptococci. *Bacteriol Rev*, 38:57-86, 1974.
12. Franciosi, R. A., J. D. Knostman, and R. A. Zimmerman. Group B

streptococcal neonatal and infant infections. *J Pediat, 82*:707-718, 1973.
13. Hall, R. T., W. Barnes, L. Krishnan, D. J. Harris, P. G. Rhodes, J. Fayez, and G. L. Miller. Antibiotic treatment of parturient women colonized with group B streptococci. *Am J Obstet Gynecol, 124*:630-634, 1976.
14. Johnson, R. H., and K. L. Vosti. Purification of two fragments of M protein from a strain of group A, type 12 streptococcus. *J Immunol, 101*:381-391, 1968.
15. Lancefield, R. C. Current knowledge of type-specific M antigens of group A streptococci. *J Immunol, 89*:307-313, 1962.
16. Lancefield, R. C. Current problems in studies of streptococci. *J Gen Microbiol, 55*:161-163, 1969.
17. Lancefield, R. C. A serological differentiation of human and other groups of hemolytic streptococci. *J Exper Med, 57*:571-595, 1933.
18. Lancefield, R. C. Serological differentiation of specific types of bovine hemolytic streptococci (group B). *J Exper Med, 59*:441-458, 1934.
19. Lancefield, R. C. Specific relationship of cell composition to biological activity of hemolytic streptococci. *The Harvey Lectures, 36*:251-290, 1940-41.
20. Lancefield, R. C. Two serological types of group B hemolytic streptococci with related but not identical, type-specific substances. *J Exper Med, 67*:25-40, 1938.
21. Lancefield R. C., and R. Hare. The serological differentiation of pathogenic and non-pathogenic strains of hemolytic streptococci from parturient women. *J Exper Med, 61*:335-349, 1935.
22. Lancefield, R. C., M. McCarty, and W. N. Everly. Multiple mouse-protective antibodies directed against group B streptococci: special reference to antibodies effective against protein antigens. *J Exper Med, 142*:165-179, 1975.
23. Maxted, W. R., J. P. Widdowson, C. A. M. Fraser, L. C. Ball, and D. C. J. Bassett. The use of the serum opacity reaction in the typing of group A streptococci. *J Med. Microbiol, 6*:83-90, 1973.
24. Morbidity and Mortality Weekly Report. Vol. 22, No. 53, 1973.
25. Ofek, I., S. Bergner-Rabinowitz, and A. M. Davies. Opsonic capacity of type-specific streptococcal antibodies. *Isr J Med Sci, 5*:293-296, 1969.
26. Quintiliani, R., and G. A. Engh. Overwhelming sepsis associated with group A beta hemolytic streptococci. *J Bone Joint Surg, 53*:1391-1399, 1971.
27. Russell, H., and R. R. Facklam. Guanidine extraction of streptococcal M protein. *Infect Immun 12*:679-686, 1975.
28. Russell, H., R. R. Facklam, and L. R. Edwards. Enzyme-linked immunosorbent assay for streptococcal M protein antibodies. *J Clin Microbiol, 3*:501-505, 1976.
29. Skjold, S. A., and L. W. Wannamaker. Method for phage typing group A type 49 streptococci. *J Clin Microbiol, 4*:232-238, 1976.
30. Steere, A. C., R. C. Aber, L. R. Warford, K. E. Murphy, J. C. Feeley, P. S.

Hayes, H. W. Wilkinson, and R. R. Facklam. Possible nosocomial transmission of group B streptococci in a newborn nursery. *J Pediat,* 87:784-787, 1975.

31. Wilkinson, H. W. Immunochemistry of purified polysaccharide type antigens of group B streptococcal types Ia, Ib, and Ic. *Infect Immun,* 11:845-852, 1975.
32. Wilkinson, H. W., and R. G. Eagan. Type-specific antigens of group B type Ic streptococci. *Infect Immun,* 4:596-604, 1971.
33. Wilkinson, H. W., R. R. Facklam, and E. C. Wortham. Distribution by serological type of group B streptococci isolated from a variety of clinical material over a five-year period (with special reference to neonatal sepsis and meningitis). *Infect Immun,* 8:228-235, 1973.
34. Wilkinson, H. W., and W. L. Jones. Radioimmunoassay for measuring antibodies specific for group B streptococcal types Ia, Ib, Ic, II, and III. *J Clin Microbiol,* 3:480-485, 1976.
35. Wilkinson, H. W., and M. D. Moody. Serological relationships of type I antigens of group B streptococci. *J Bacteriol,* 97:629-634, 1969.
36. Wilson, A. T. The relative importance of the capsule and the M-antigen in determining colony form of group A streptococci. *J Exper Med,* 109:257-270, 1959.
37. Zahradnicky, J., M. Lukesova, R. Kronus, J Jelinkova, V. Hejda, M. Valchova, and S. Turkova. An unusual case of infection caused by a group B streptococcus. *J Hyg Epidem Microbiol Immunol,* 16:21-27, 1972.

Chapter 6

BIOTYPING OF ANAEROBIC BACTERIA ASSOCIATED WITH HUMAN DISEASE

V. R. Dowell, Jr., G. L. Lombard, M. D. Stargel,
S. D. Allen, F. S. Thompson, A. Y. Armfield

Conventional and micro-techniques were used for biogrouping and biotyping a variety of anaerobic bacteria, and results were compared. Included in this report are data illustrating the value of biogrouping isolates from clinical specimens, data on *Bacteroides* and *Fusobacterium* isolates from patients with bacteremia, and data obtained when micromethods were used for biotyping *Bacteroides fragilis* subspecies *fragilis* and *Propionibacterium acnes*.

Also, data from the examination of blood culture isolates of *P. acnes* from a patient with endocarditis with a micromethod (API Lactobacillus System) are presented to illustrate a specific use of biotyping.

The information derived from biogrouping of anaerobic bacteria was found to be very useful in studying the role of these microorganisms in disease. Some biogroups are commonly associated with disease and others are not. Using the API Lactobacillus 50 System, four biogroups of *Bacteroides fragilis* ss *fragilis* and fourteen biotypes of *P. acnes* were recognized. Six blood culture isolates from the patient with endocarditis were found to be identical, biotype e.

BIOTYPING of bacteria, i.e. determining the enzymatic capabilities of the microorganisms by performing physiological and biochemical tests, is a powerful

---
*We wish to thank Ms. Jana Swenson, Antimicrobics Investigations Section, for performing the antibiotic susceptibility tests and Ms. Joy Maxey for her assistance with the biotyping of *P. acnes* cultures.

Use of trade names is for identification only and does not constitute endorsement by the Public Health Service or by the U. S. Department of Health, Education and Welfare.

tool in clinical microbiology. This technique is very useful for the accurate identification of bacteria, for assessing the clinical significance of bacterial isolates, and for epidemiologic investigations. In this presentation, we describe conventional and microtechniques for biogrouping anaerobic bacteria, and we provide data to illustrate the potential value of biotyping in assessing the clinical significance of anaerobe isolates associated with human disease.

Biogrouping Bacteroidaceae

In the classification of anaerobic bacteria as outlined in the seventh edition of *Bergey's Manual of Determinative Bacteriology*,[2] the use of morphologic features for primary differentiation was heavily relied upon. However, because of a number of pitfalls in this system, it became apparent that other tests were needed for accurate characterization of anaerobes.

For this reason, the use of a battery of cultural and biochemical tests (Table 6-I) was initiated during the 1960s for the examination of all anaerobe isolates submitted to the Center for Disease Control (CDC) Anaerobe Laboratory for identification or confirmation. The results of these tests — along with data obtained from the examination of reference cultures with the same tests — were then tabulated, and differential tables were compiled in 1968. These data indicated that the majority of saccharolytic bacteroides isolates received from state health department and federal laboratories in the United States could be separated into the following five groups: *Bacteroides fragilis, Bacteroides incommunis, Bacteroides oralis, Bacteroides variabilis,* and *Bacteroides terebrans.* At that time, it was emphasized that the species names and group designations used in the tables were tentative, pending the adoption of acceptable nomenclature by taxonomists.[5]

Soon thereafter Moore and his colleagues at the Virginia Polytechnic Institute (VPI) reported the use of metabolic product analysis by gas liquid chromatography (GLC) and described various other differential tests for the characterization of anaerobes in which pre-reduced anaerobically sterilized (PRAS)

Table 6-I. BACTERIOLOGIC TESTS USED BY THE CENTER
FOR DISEASE CONTROL FOR DIFFERENTIATION
OF ANAEROBES, 1968*

1. Examination of microscopic and colonial characteristics
2. Observation of growth in liquid media, motility, and tolerance to oxygen
3. Tests for fermentation of glucose, mannitol, lactose, sucrose, maltose, salicin, glycerol, fructose, xylose, and arabinose
4. Tests for hydrolysis of starch, gelatin; and esculin
5. Tests for production of lecithinase, lipase, catalase, H_2S, indole, and urease
6. Tests for reduction of nitrate, action in milk, action on meat particles, hemolysis of rabbit blood, gas production in glucose broth, and motility in motility medium

*Adapted from Dowell and Hawkins[6]

media were used.[1] They also proposed various changes in the classification of the nonsporeforming anaerobes, many of which were subsequently included in the eighth edition of *Bergey's Manual of Determinative Bacteriology*.[3] The saccharolytic bacteroides listed in the 1970 VPI anaerobe Manual[1] are shown in Table 6-II, and the key characteristics which were proposed for differentiating the five subspecies of *B. fragilis* are presented in Table 6-III.

During 1972, we began using a new battery of tests for characterizing anaerobes in the CDC Anaerobe Laboratory to allow for subspeciation of *B. fragilis,* and we used these tests to re-examine fifty previously identified saccharolytic bacteroides (Table 6-IV). The major differences between the new and old data were that approximately 10 percent of the strains previously identified as *B. fragilis* were found to be *B. fragilis* ss *distasonis* and that other strains which we had initially called *B. oralis* were identified as *B. fragilis. B. fragilis* ss *ovatus* was conspicuously absent from the clinical isolates submitted to CDC for identification during the period April 1971 to March 1972 (Table 6-V). To date (April 1976), we have not isolated *B. fragilis* subspecies *ovatus* from a properly collected clinical specimen of a normally sterile area of the body. However, this subspecies appears to reside in large numbers in the human

Table 6-II. SACCHAROLYTIC BACTEROIDES IDENTIFIED BY
THE VPI ANAEROBE LABORATORY, 1970*

Bacteroides fragilis

    ss *fragilis*
    ss *distasonis*
    ss *ovatus*
    ss *thetaiotaomicron*
    ss *vulgatus*

Bacteroides oralis

    ss *oralis*
    ss *elongatus*

Bacteroides trichoides
Bacteroides biacutus

*VPI anaerobe manual[1]

Table 6-III. KEY CHARACTERISTICS FOR DIFFERENTIATING
THE SUBSPECIES OF BACTEROIDES FRAGILIS*

Subspecies	Indole	Mannitol	Trehalose	Rhamnose
fragilis	−	−	−	−
distasonis	−	−	+	V
ovatus	+	+	+	+
thetaiotaomicron	+	−	V	+
vulgatus	−	−	−	+

*Adapted from the VPI anaerobe manual[1]

intestinal tract as part of the normal microbiota.[11,12,14] *B. fragilis* ss *fragilis* is by far the most common anaerobic, non-spore-forming gram-negative rod isolated from clinical specimens[7,14] although it may be outnumbered by other subspecies of *B. fragilis* in the normal intestinal tract. This suggests a fundamental pathogenic difference between subspecies *fragilis* and the other subspecies of *B. fragilis*.

Table 6-IV. RESULTS OF THE REEXAMINATION OF 50 SACCHAROLYTIC BACTEROIDES 1972

Number of Strains	Former Designation	New Designation	Number of Strains
35	*B. fragilis*	*B. fragilis* ss *fragilis*	32
		ss *distasonis*	3
2	*B. incommunis*	*B. fragilis* ss *vulgatus*	2
6	*B. variabilis*	*B. fragilis* ss *thetaiotaomicron*	6
2	*B. terebrans*	*B. trichoides**	2
5	*B. oralis*	*B. oralis*	1
		B. fragilis ss *fragilis*	4

*Subsequently identified as *Clostridium ramosum*[13]

The organism listed as *Bacteroides trichoides* in Table 6-IV — as well as those formerly called *Bacteroides terebrans, Catenabacterium filamentosum,* and *Eubacterium filamentosum* — are now classified as *Clostridium ramosum*.[13] Some strains of this species are easily confused with bacteroides because of their tendency to stain gram-negative and their failure to produce spores on media routinely used for culturing anaerobes.[7,14]

Bateroides Bacteremia

The value of biogrouping anaerobes was evident in our 1971

Table 6-V. DISTRIBUTION OF *B. FRAGILIS* SUBSPECIES* IN THE CLINICAL ISOLATES OF SACCHAROLYTIC BACTEROIDES RECEIVED BY CDC FROM APRIL 1971- MARCH 1972

Subspecies	Number
B. fragilis ss *fragilis*	49
ss *thetaiotaomicron*	10
ss *distasonis*	6
ss *vulgatus*	5
ss *ovatus*	0

*Subspecies identified as described by Moore and Holdeman[1]

study of clinical and bacteriological data from 250 patients with bacteremia caused by anaerobic, non-spore-forming, gram-negative bacilli.[10] The blood culture isolates were subjected to a variety of bacteriologic tests: fermentation of glucose, mannitol, lactose, maltose, xylose, and arabinose; hydrolysis of gelatin and esculin; nitrate reduction; action in milk; gas production in glucose broth; production of H_2S, indole, and urease; and production of volatile acids, as detected by gas liquid chromatography, in peptone yeast extract glucose (PYG) broth.[1,7] On the basis of these tests, ten tentative species or biogroups of *Bacteroides* and five species or biogroups of *Fusobacterium* were identified (Table 6-VI). We found that consideration of each *Bacteroides* and *Fusobacterium* species or group separately, rather than grouping them together as others had done in the past, was paramount to a clearer understanding of their role in human disease. Support for this approach to the investigation of anaerobic bacterial infection with bacteremia was soon provided by the reports of others.[4,22,26]

Table 6-VI. BACTEROIDES AND FUSOBACTERIUM SPECIES AND GROUPS IDENTIFIED IN A STUDY OF 250 PATIENTS WITH BACTEREMIA BY FELNER AND DOWELL (10)

Designation (1971)	Present Designation (1976)	Number of Patients
B. fragilis	B. fragilis ss fragilis and ss distasonis*	140
B. oralis	B. oralis and B. fragilis ss fragilis**	13
B. variabilis	B. fragilis ss thetaiotaomicron	11
B. incommunis	B. fragilis ss vulgatus	8
B. terebrans	Clostridium ramosum	4
B. melaninogenicus	B. melaninogenicus ss asaccharolyticus and ss intermedius**	4
B. corrodens	B. corrodens	1
CDC group F-1	CDC group F-1	10
CDC group F-3	CDC group F-3	2
Unspeciated Bacteroides	Unspeciated Bacteroides	2
F. fusiforme	F. nucleatum	21
F. necrophorum	F. necrophorum	20
F. ridiculosum	F. mortiferum	8
F. girans	Clostridium species	1
Unspeciated Fusobacterium	Unspeciated Fusobacterium	5

*Approximately 10% of those previously identified as B. fragilis were B. fragilis ss distasonis. Therefore, it is estimated that 126 of the B. fragilis were B. fragilis ss fragilis and 14 ss distasonis.
**The number of each species or subspecies was not determined.

The distribution of the subspecies of *B. fragilis* in 104 isolates received at CDC since 1971 from 100 patients with bacteremia is shown in Table 6-VII. Detailed clinical and bacteriological data on patients with Bacteroides and Fusobacterium bacteremia which we have investigated in recent years will be presented in a separate report.

Table 6-VII. DISTRIBUTION OF SUBSPECIES IN 104 ISOLATES FROM 100 PATIENTS WITH *BACTEROIDES FRAGILIS* BACTEREMIA EXAMINED BY THE CDC ANAEROBE SECTION FROM 1971-1975

Subspecies	Number	Percent
fragilis	73	70.3
thetaiotaomicron	19	18.2
vulgatus	7	6.7
distasonis	5	4.8
ovatus	0	0.0

Use of Micromethods for Biotyping

During the last few years, investigators in our laboratory have worked on the improvement and evaluation of commercially available microsystems for use in the biochemical characterization of anaerobic bacteria.[24,23] On the basis of this experience, we feel that the future of micromethods in facilitating the accurate identification and biotyping of anaerobes is quite promising. So far, the API20, the Minitek, and the API Lactobacillus fifty systems have been tested.

API 20 System

During 1973, we evaluated the API 20 micromethod system

(Analytab Products, Inc., New York, NY) for use in the biochemical characterization of anaerobic bacteria.[24] In a study of 104 cultures, results of 91 percent of the tests performed with the API system were in agreement with those obtained by conventional methods. Only five of the seventeen differential tests compared gave less than 90 percent agreement. From this study, we concluded that substituting rapid, economical, microbiochemical techniques for some of the more expensive, time-consuming conventional tests for identification of anaerobes is quite feasible. On the basis of our study results — and with the advice of others — the Analytab Corporation prepared a modified API 20 Anaerobe System, which is now marketed commercially. Two years later, Moore, Sutter, and Finegold[17] reported that the modified API system maintained a satisfactory performance level in the characterization of anaerobes in their laboratory.

Minitek System (BBL)

We recently[23] reported results of a study on the usefulness of the Minitek Miniaturized Microorganism Differentiation System, marketed by BBL (Cockeysville MD) for characterization of anaerobes. This system consists of paper disks impregnated with appropriate substrates which are dispensed into the wells of a special Minitek plastic plate. A cell suspension of the test organism is inoculated into each of the wells with a pipetting device specially designed to accommodate disposable pipettes. Only 0.05ml of cell suspension is required for each well. After the inoculum is added to the substrate disks, the plates are incubated under increased humidity for an appropriate time, and the reactions are read and recorded.

In our study the Minitek System was modified for characterization of anaerobes by using a new suspension medium (Lombard-Dowell broth) and by increasing the number of bacteria in the inoculum. The modified Minitek System and the conventional CDC method were used to test a variety of anaerobic bacteria, and results were compared. Tests performed by both techniques were indole and H_2S production, esculin hydrolysis, nitrate reduction, and fermentation of glucose, mannitol, lactose, sucrose, maltose, salicin, glycerol, xylose, arabinose, mannose, rhamnose, and trehalose. The Minitek results, recorded after forty-eight hours, agreed satisfactorily with the conventional test results, which were usually recorded after five to seven days incubation.[7] In the examination of eighty strains representing twenty-two different species or subspecies of anaerobic bacteria with sixteen biochemical tests performed in triplicate, 93.8 percent of the Minitek results agreed with those of the corresponding conventional tests. Only the tests for indole, H_2S, and nitrate reduction gave less than 90 percent agreement. Except for use with these three tests, we concluded that the Modified Minitek System is a suitable substitute for the more expensive and time-consuming conventional procedure.

Lactobacillus 50 System

More recently, we have tested the API Lactobacillus 50 System for use in biotyping anaerobic bacteria. This system, developed for characterization of lactobacilli and related organisms,[18] is similar to the API 20 Anaerobe System, but contains fifty substrates in plastic wells rather than twenty. We have used the system for biotyping anaerobic bacteria as follows:

a. A tube of Lombard-Dowell (LD) broth[23] is inoculated from a well-isolated single colony on blood agar and is incubated anaerobically at 35C until definite turbidity is evident (usually within 18 hours).
b. The purity of the broth culture is checked by examining a gram-stained smear of the cells.
c. If the culture appears to be pure, a pasteur pipette is used to fill each of the fifty cupules of the system with bacterial suspension. The strips are placed in a special plastic humidor provided by the manufacturer and are incubated at 35°C for 48 h in an anaerobic glove box[7] containing an atmosphere of approximately 5 percent CO_2, 10 percent H_2, and 85 percent N_2. Final readings are then made.

Biotyping Bacteroides fragilis ss fragilis

The composite data obtained from the examination of 66 *B. fragilis* ss *fragilis* strains with the API Lactobacillus 50 System are shown in Table 6-VIII. The reactions of the *B. fragilis* ss *fragilis* strains, which included isolates from hospitals throughout the United States, were remarkably consistent in some of the substrates. However, four of the substrates (D arabinose, L arabinose, D xylose, L xylose) allowed separation of the strains into four distinct biotypes (Table 6-IX). Such data should be very useful in future studies to determine the role of anaerobic microorganisms in health and disease.

Biotyping Propionibacterium acnes

Because of their ubiquity, the propionibacteria have often been regarded as laboratory contaminants. These microorganisms constitute a part of the microflora of human skin, where they reside in sebaceous glands and in hair follicles.[19] They appear to be constantly shed from skin and hair and tend to survive exposure and drying.[27] The propionibacteria are also commonly isolated from a variety of habitats lined by mucous membranes, including the mouth,[21] nasal cavity,[25] and gastrointestinal tract.[21] These habitats are strikingly similar to those of the staphylococci, which likewise may or may not be contaminants. It seems as likely that a change in the integrity of skin or mucous membrane surfaces would provide portals of entry for *P. acnes* just as for *Staphylococcus aureus*.

Perhaps because of preconceived notions that "anaerobic diptheroids" are nonpathogens, infections caused by *P. acnes* frequently go undiagnosed. They may also be missed because of inadequacies in specimen handling, poor anaerobic technique, or incorrect identification of clinical isolates. Evidence is mounting, however, that *P. acnes* may sometimes play a significant role in the pathogenesis of certain human infections. Probably the strongest evidence for this has accumulated in case studies of bacterial endocarditis, especially in patients with chronic valvular disease or cardiac valve prostheses.

Table 6-VIII. REACTIONS OF SIXTY-SIX *BACTEROIDES FRAGILIS* SS *FRAGILIS* STRAINS IN THE API LACTOBACILLUS 50 SYSTEM*

Test or Substrate	% +	Test or Substrate	% +
N-Acetyl-glucosamine	100	Methyl-D-Mannoside	18.2
Esculin	100	Inulin	12.2
D-Fructose	100	Amylose	1.5
Glucose	100	D-Melezitose	1.5
D-glucose	100	Adonitol	0
Lactose	100	L-Arabinose	0
Maltose	100	Arbutin	0
D-Mannose	100	Arginine	0
D-Raffinose	100	Dulcitol	0
Sucrose	100	Erythritol	0
Galactose	98.5	Glycerol	0
Glycogen	98.5	M-Inositol	0
Starch	98.5	Mannitol	0
Dextrin	97.0	Methyl xyloside	0
D-Arabinose	92.4	Nitrate	0
D-Melibiose	90.9	Rhamnose	0
Amygdalin	89.4	Salicin	0
D-Xylose	84.8	Sorbitol	0
D-Cellobiose	81.8	L-Sorbose	0
Methyl-D-glucoside	69.7	D-Trehalose	0
Ribose	30.3	L-Xylose	0

*Reactions obtained with 7 substrates of the system (Teepol, 0.4%; Teepol, 0.6%; NaCl, 4%; NaCl, 6%; NaCl, 10%; ONPG; Pyruvic acid and [V.P.]) are not included.

Table 6-IX. BIOGROUPS OF *B. FRAGILIS* SS *FRAGILIS* DETECTED
WITH THE API LACTOBACILLUS 50 SYSTEM

Biogroup	Number of strains	Percent	Fermentation of			
			D-arabinose	L-arabinose	D-xylose	L-xylose
1	54	81.8	+	−	+	−
2	7	10.6	+	−	−	−
3	2	3.0	−	−	+	−
4	3	4.5	−	−	−	−

In reviewing the clinical records of patients with endocarditis, from whom cultures had been submitted to CDC (from 1963 through 1969), Felner and Dowell[9] reported that, in five of thirty-three cases of anaerobic bacterial endocarditis, *P. acnes* was the predominant or only organism. In a later study, *P. acnes* was the predominant organism in fifteen out of forty-eight cases of anaerobic bacterial endocarditis.[8] Patients in that series ranged from five to seventy-eight years, but 65 percent were more than thirty-nine years old. Probable portals of entry for *P. acnes* were the oropharynx (2 cases), gastrointestinal tract (2 cases), and skin (3 cases), but for seven of the patients the portals were unknown. The clinical findings were generally similar to those of patients with endocarditis resulting from facultative anaerobes. Five of the fifteen patients with *P. acnes* endocarditis had undergone cardiovascular surgical procedures, and cardiac valve prostheses had been inserted in three. This series in no way reflects the true incidence of *P. acnes* endocarditis, which is probably rare. However, as Felner pointed out, there are several additional reports in the literature which suggest that *P. acnes* endocarditis is not infrequent in patients who have undergone open heart surgery.[15,16] Several such patients had endocarditis after insertion of Starr-Edwards ball-valve prostheses. The pathogenesis is unknown. The organisms might be massively seeded into the blood stream by the surgeon during the extensive skin incision, or might contaminate suture material or other foreign materials introduced into the patient. The cardiopulmonary bypass machinery is yet another potential source of contamination. If the organisms attach either to the prosthetic device or to the valve surface, and metabolize slowly in a matrix of fibrin and other plasma protein material, the bacteria could be relatively protected from and resistant to the usual prophylactic antibiotics. Besides the purely cardiac complications of such procedures, bacteremia with all its complications (such as metastatic abscesses) might occur. The final outcome was death in five patients of Felner's series.[8]

In recent months, we have explored the possibility of using biotyping as an aid in assessing the clinical significance of *P. acnes* isolates from clinical materials associated with disease and in determining, if possible, the portal of entry and source of the microorganism involved.

The composite reactions obtained from the examination of 675 *P. acnes* isolates from clinical samples with conventional tests currently used by the CDC Anaerobe Section[7] are listed in Table 6-X. The more consistent characteristics of the organism, such as fermentation of glucose (100%), fermentation of glycerol (99%), catalase production (99%), propionic acid production (100%), etc., are quite evident from these data.

Data obtained from testing sixty-six cultures of *P. acnes* with the API Lactobacillus 50 System are shown in Table 6-XI. These included seventeen strains from G. Pulverer, Cologne, Germany,[20] and forty-nine isolates from various sources submitted by state health department and federal laboratories in the United States. Using data from the fermentation of five substrates (adonitol, erythritol, fructose, galactose and sorbitol), we were able to distinguish fourteen biotypes of *P. acnes* (Table 6-XII). Using a different combination of substrates, Pulverer and Ko were able to distinguish eight biotypes of the organism.[20]

Table 6-X. COMPOSITE REACTIONS OF 675 *P. ACNES* WITH VARIOUS DIFFERENTIAL TESTS*

Test	Percent Positive
Fermentation of:	
L-arabinose	0
Dextrose	100
Glycerol	99
Lactose	1
Maltose	1
Mannitol	28
Mannose	95
Rhamnose	0
Salicin	0
Starch	1
Sucrose	1
Trehalose	9
D-xylose	0
Aerotolerance	8
Catalase production	99
Esculin hydrolysis	1
Gelatin hydrolysis	74
Hemolysis (rabbit blood)	45
H_2S production	3
Indol production	81
Action on milk	Clot 92, digestion 1
Motility	0
Nitrate reduction	94
Starch hydrolysis	1
Urease production	0
Metabolic products:	
Acetic	100
Propionic	100
Isobutyric	0
Butyric	0
Isovaleric	22
Valeric	0
Isocaproic	0
Caproic	0
Lactic	45
Succinic	0

*Tests performed as described in Dowell and Hawkins[7]

Table 6-XI. REACTIONS OF 66 *PROPIONIBACTERIUM ACNES* CULTURES IN THE API LACTOBACILLUS 50 SYSTEM*

Test or Substrate	% +	Test or Substrate	% +
Glucose	100	L-xylose	0
Glycerol	100	Methyl xyloside	0
D-mannose	100	L-sorbose	0
Ribose	95	Rhamnose	0
Nitrate	94	Dulcitol	0
D-Fructose	88	Methyl-D-mannoside	0
N-Acetyl-glucosamine	88	Methyl-D-glucoside	0
Galactose	83	Amygdalin	0
Sorbitol	59	Arbutin	0
Adonitol	45	Esculin	0
Erythritol	42	Salicin	0
Trehalose	9	D-cellobiose	0
Sucrose	6	Lactose	0
Maltose	5	D-melibiose	0
Inositol	5	Inulin	0
Mannitol	4	D-raffinose	0
Melezitose	3	Dextrin	0
D-arabinose	2	Amylose	0
L-arabinose	2	Starch	0
D-xylose	0	Glycogen	0

*Reactions obtained with Teepol, 0.4%; Teepol, 0.6%; NaCl, 4%; NaCl, 6%; NaCl, 10%; ONPG; pyruvic acid and (V.P.) are not included.

Propionibacterium acnes Endocarditis

We recently were consulted concerning a sixty-eight-year-old woman with a diagnosis of endocarditis, from whom five of six blood cultures (two bottles each culture, thioglycollate and Trypticase Soy Broth) had yielded *P. acnes*. Six positive bottles from three different blood cultures were received by our laboratory for examination. The isolates from each were identified as *P. acnes* with our battery of conventional tests and gas liquid chromatography,[7] and the fermentation patterns of the six isolates obtained with the API Lactobacillus System were identical (biotype e, Table 6-XII), as shown in Table 6-XIII. The

Table 6-XII. BIOTYPES OF PROPIONIBACTERIUM ACNES* DISTINGUISHED WITH THE API LACTOBACILLUS 50 SYSTEM

Biotype	Number of Strains	Fermentation of				
		Adonitol	Erythritol	Galactose	Fructose	Sorbitol
a	2	−	−	−	−	−
b	3	−	−	+	−	−
c	4	−	−	−	+	−
d	4	−	−	+	+	−
e	5	+	+	+	+	+
f	5	+	−	+	+	+
g	15	−	−	+	+	+
h	4	+	−	+	+	−
i	7	+	+	+	+	−
j	3	−	+	+	+	+
k	3	−	+	+	+	−
l	1	−	+	−	−	−
m	1	+	−	−	+	−
n	1	+	+	+	−	+

*58 strains examined

minimal inhibitory concentration (MIC) antibiotic susceptibility patterns of the isolates were also essentially identical (Table 6-XIV). On the basis of these data, it appears that each of the six positive blood culture bottles contained the same strain of *P. acnes,* and this strain probably was responsible for the patient's endocarditis.

Table 6-XIII. FERMENTATION REACTIONS* OF SIX BLOOD CULTURE ISOLATES OF *P. ACNES* FROM A PATIENT WITH ENDOCARDITIS, IN TESTS WITH API LACTOBACILLUS 50 SYSTEM

Blood Culture Isolate	Fermentation of				
	Adonitol	Erythritol	Galactose	Fructose	Sorbitol
1	+	+	+	+	+
2	+	+	+	+	+
3	+	+	+	+	+
4	+	+	+	+	+
5	+	+	+	+	+
6	+	+	+	+	+

*Fermentation reactions are consistent with biotype e, Table 6-XII.

Table 6-XIV. ANTIBIOTIC SUSCEPTIBILITY* (MINIMAL INHIBITORY CONCENTRATION G/ML) OF *P. ACNES* BLOOD CULTURE ISOLATES FROM PATIENT WITH ENDOCARDITIS

Isolate	Gentamicin	Penicillin	Tetracycline	Cephalothin	Cloramphenicol	Erythromycin	Clindamycin	Lincomycin
1	1.0	≤ 0.06	0.25	≤ 0.06	0.25	≤ 0.06	≤ 0.06	≤ 0.06
2	1.0	≤ 0.06	0.25	≤ 0.06	0.25	≤ 0.06	≤ 0.06	≤ 0.06
3	1.0	≤ 0.06	0.50	≤ 0.06	0.50	≤ 0.06	≤ 0.06	≤ 0.06
4	1.0	≤ 0.06	0.50	≤ 0.06	0.50	≤ 0.06	≤ 0.06	≤ 0.06
5	1.0	≤ 0.06	0.25	≤ 0.06	0.25	≤ 0.06	≤ 0.06	≤ 0.06
6	1.0	≤ 0.06	0.50	≤ 0.06	0.25	≤ 0.06	≤ 0.06	≤ 0.125

*Antibiotic susceptibility tests performed by the Antimicrobics Investigation Section, Clinical Bacteriology Branch, Bacteriology Division, Center for Disease Control.

REFERENCES

1. Anaerobe Laboratory, Virginia Polytechnic Institute. *Outline of Clinical Methods in Anaerobic Bacteriology*, VPI, Blacksburg, Virginia, 1970.
2. Breed, R. S., E. G. D. Murray, and N. R. Smith. *Bergey's Manual of Determinative Bacteriology* (7th Ed.), Baltimore, William & Wilkins Co., 1957.
3. Buchanan, R. E. and N. E. Gibbons (Eds.), *Bergey's Manual of Determinative Bacteriology*, Eighth Edition. Baltimore, Williams & Wilkins Co., 1974.
4. Chow, A. W. and L. B. Guze. Bacteroidaceae bacteremia: clinical experience with 112 patients. *Medicine, 53*:93-126, 1974.
5. Dowell, V. R., Jr. Anaerobic Infections. Chapter XV in *Diagnostic Procedures for Bacterial, Mycotic and Parasitic Infections*, Fifth Edition. Edited by H. L. Bodily, E. L. Updyke, J. O. Mason. American Public Health Association, (Pages 494-543), 1970.
6. Dowell, V. R., Jr. and T. M. Hawkins. *Laboratory Methods in Anaerobic Bacteriology, CDC Laboratory Manual*. U. S. Department of Health, Education, and Welfare, Public Health Service, Pub. No. 1803, June 1968.
7. Dowell, V. R., Jr., and T. M. Hawkins. *Laboratory Methods in Anaerobic Bacteriolgoy, CDC Laboratory Manual*. U. S. Department of Health, Education, and Welfare, Public Health Service, Atlanta, Center for Disease Control, DHEW Pub. No. (CDC) 74-8272, 1974.
8. Felner, J. M. Infective endocarditis caused by anaerobic bacteria. In *Anaerobic Bacteria Role In Disease*. Edited by A. Balows, R. M. DeHaan, V. R. Dowell, Jr., and L. B. Guze. Springfield, Thomas, pp. 345-352, 1974.
9. Felner, J. M. and V. R. Dowell, Jr. Anaerobic bacterial endocarditis. *New Engl J Med, 283*:1188-1192, 1970.
10. Felner, J. M. and V. R. Dowell, Jr. "Bacteroides bacteremia." *Am J Med, 50*:787-796, 1971.
11. Finegold, S. M., H. R. Attebery, and V. L. Sutter. Effect of diet on human fecal flora: comparison of Japanese and American Diets. *Am J Clin Nutr, 27*:1456-1469, 1974.
12. Finegold, S. M., D. J. Flora, H. R. Attebery and V. L. Sutter. Fecal bacteriology of colonic polyp patients and control patients. *Cancer Research, 35*:3407-3417, 1975.
13. Holdeman, L. V., E. P. Cato and W. E. C. Moore. *Clostridium ramosum* (Veillon and Zuber) *comb. nov.*: Emended description and proposed neotype strain. *Int J Syst Bacteriol, 21*:35-39, 1971.
14. Holdeman, L. V., E. P. Cato, and W. E. C. Moore. Current classification of clinically important anaerobes. Chapter VIII in *Anaerobic Bacteria Role in Disease*. Edited by A. Balows, R. M. DeHaan, V. R. Dowell, Jr., and L. B. Guze. Springfield, Thomas, 1974. pp. 67-74.

15. Johnson, W. D., C. G. Cobbs, L. J. Anditi and D. Kaye. Diphtheroid endocarditis after insertion of a prosthetic heart valve. *JAMA, 203*:117-119, 1968.
16. Levin, J. Diphtheroid bacterial endocarditis after insertion of a Starr valve. *Ann Int Med, 64*:396-398, 1965.
17. Moore, H. B., V. L. Sutter and S. M. Finegold. Comparison of three procedures for biochemical testing of anaerobic bacteria. *J Clin Microbiol, 1*:15-24, 1975.
18. Paule, R. Contribution à l étude biochimique du genre Lactobacillus par une méthode normalisée These doctorate en Pharmacie, Lyon, 14 Juin 1971.
19. Puhvel, S. M. and R. M. Reiser. Dermatologic anaerobic infections. Chapter 34. In *Anaerobic Bacteria Role in Disease*. Edited by A. Balows, R. M. DeHaan, V. R. Dowell, Jr. and L. B. Guze. Springfield, Thomas, 1974, 435-450.
20. Pulverer, G. and H. L. Ko. Fermentative and serologic studies on *Propionibacterium acnes*. *Appl Microbiol, 25*:222-229, 1973.
21. Smith, L. DS. *The Pathogenic Anaerobic Bacteria*. 2nd Ed., Springfield, Thomas, 1975, 68-73.
22. Sonnenwirth, A. C. Incidence of intestinal anaerobes in blood cultures. Chapter XV. In *Anaerobic Bacteria Role in Disease*. Edited by A. Balows, R. M. DeHaan, V. R. Dowell, Jr. and L. B. Guze. Springfield, Thomas, 1974, 157-171.
23. Stargel, M. D., F. S. Thompson, S. E. Phillips, G. L. Lombard and V. R. Dowell, Jr. Modification of the Minitek miniaturized differentiation system for characterization of anaerobes. *J Clin Microbiol, 3*:291-301, 1976.
24. Starr, S. E., F. S. Thompson, V. R. Dowell, Jr. and A. Balows. Micromethod system for identification of anaerobic bacteria. *Appl Microbiol, 25*:713-717, 1973.
25. Watson, E. D., N. J. Hoffman, R. W. Simmers and T. Rosebury. Aerobic and anaerobic bacterial counts of nasal washings: presence of organisms resembling *Corynebacterium acnes*. *J Bacteriol, 83*:144, 1962.
26. Wilson, W. R., W. J. Martin, C. J. Wilhowshe and J. A. Washington. Anaerobic bacteremia. *Mayo Clinic Proc, 47*:639-646, 1972.
27. Woodroffe, R. C. S. and D. A. Shaw. Natural control and ecology of microbial populations on skin and hair. In *The Normal Microbial Flora of Man*. Edited by F. A. Skinner and J. G. Carr. New York, Academic Press, 1974, pp. 13-34.

Chapter 7

STANDARDIZATION AND AUTOMATION IN BIOTYPING*

J. J. FARMER III

BIOTYPING is becoming one of the most useful methods for "fingerprinting" bacteria. Because many species can be divided into useful "biotypes" with simple laboratory tests, biotyping is feasible even in a small laboratory. Since ten to fifty biochemical reactions used in biotyping can be tested with commercially available systems,[1,20,23,26] biotyping should become one of the most useful methods for speciation and subspeciation.

Biotyping has many important applications. In making taxonomic changes, it is very important to know the exact biotype involved. For example, it may be desirable to move Biotypes 5a and 5b of *Proteus rettgeri*[24] to *Providencia stuartii*. Similarly the deoxyribonuclease$^+$, sorbitol$^-$, a yellow pigmented "biotype" of *Enterobacter cloacae* is really not a biotype, but a distinct species which will probably be placed in *Enterobacter*, and "rhamnose$^+$ *Yersinia enterocolitica*" will probably be elevated from biotype to species status.[3] Biotypes often have important relationships in human disease or epidemiology. Two biotypes of *Salmonella typhimurium* that cause human infections in Canada can be traced to ducks (or geese) and to poultry, respectively, because of their specific biotype.† Rare biotypes, such as lysine$^-$ *Salmonella typhi*,[8] lactose$^+$ *S. typhimurium*,[9] or indole$^+$ *Klebsiella pneumoniae*[25] are excellent markers in tracing the epidemiology of these pathogens in

---

*I thank my colleagues in the Enteris Section, CDC, for their participation in the many different studies described in this paper.

   Use of trade names is for identification only and does not constitute endorsement by the Public Health Service or by the U. S. Department of Health, Education, and Welfare.

†Margaret Finlayson, personal communication.

family, community, or hospital outbreaks. Clinical microbiologists are beginning to use biotypes and antibiograms as the primary "epidemiological markers" in detecting and following hospital-acquired infections. As the use of biotyping increases, variables in the biochemical tests must be understood and eliminated. For this reason I shall focus on methodology in biotyping — particularly test standardization and automation.[2,7,12,15,29,31] First, however, I shall discuss the word "biotype."

For many years, the word "biotype" has been used to define members of the same species which differ in physiological properties or biochemical reactions. Terms such as "biotype," "antibiogram," "phage type," or "pathotype" are not covered by the Rules in the 1976 revision of the *International Code of Nomenclature of Bacteria*.[22] Although these infra-subspecific subdivisions (names below the species level) are not covered in the actual rules, the applications of the definitions are clarified in Appendix 10. The Code recommended that the term "biotype" be replaced by "biovar" (biochemical variety) in scientific writing. Appendix 10 states:

> the introduction of the suffix — "var"... to replace "-type" is recommended to avoid confusion with the strict use of the term "type" to mean nomenclatural type (see Rule 15). The term type in bacteriology should be used strictly for a nomenclatural type (Principal 5, and Chapter 3, Section 4). It should not be used to designate a division of a species (phage type, biotype) nor to designate taxa base on antigenic characters (serotype).

The term "biogroup" could be used, but might be confusing because, Appendix 10 states:

> The term "group" is informal and has no nomenclatural standing. It may prove useful to designate informally a set of organisms having certain characteristics in common, provided that it is used with care and exact definitions to avoid ambiguity. It should not be used to avoid the use of the correct name of a taxon such as a genus or species. However, it may be used when the bacteriologist does not wish to give a formal name to a set of bacteria until further studies have been made but wishes to publish his results and seek the

opinion of others. *Example:* IID group, later called *"Cardiobacterium hominis."*

Most microbiologists will probably prefer the term "biotype" to "biovar" because they are more familiar with "biotype." I have always used "biotype" and prefer it to "biovar"; however, the rules of nomenclature make it clear that you cannot reject a name simply because you do not like it. Thus, from this point on I will use "biovar" instead of "biotype," as suggested by Appendix 10 of the *Bacteriological Code*.[22] Presumably, the "ing" form of verb will not be replaced by "biovaring." Thus, I will continue to use "biotyping," but the results of biotyping will be "biovars" not "biotypes."

Materials and Methods

Media

The media used in this study have been described in detail by Ewing and Davis.[7] Whenever possible dehydrated media from commercial sources were used, and all unless stated otherwise incubations were at $36C \pm 1C$. Bacteria were from the collection at the Enteric Section, Center for Disease Control (CDC).

Description of Tests

Six different works were used in determining variations in biochemical tests and procedures. They were: (1) Edwards and Ewing: *Identification of Enterobacteriaceae,* third edition[6]; (2) F. Kauffman: *The Bacteriology of Enterobacteriaceae*[21]; (3) Cowan and Steel: *Identification of Medical Bacteria*[4]; (4) V. B. D. Skerman: A Guide to the *Identification of the Genera of Bacteria*[28] (5) Topley and Wilson: *Principles of Bacteriology and Immunology*[32]; and (6) *Manual of Microbiological Methods* of the Society of American Bacteriologists.[30] Other well-known manuals were not used because the authors of the pertinent chapter(s) were represented in the works cited. The descriptions of the tests were compared for possible variation in media, inoculation, incubation, or interpretation.

Color Standards

A set of color standards was commercially available for bromthymol blue, methyl red, and phenyl red.* The standards show the color of the indicator at different pHs (Fig. 7-1). For example, the color standards for bromthymol blue contained 9 ampules from pH 6.0 to 7.6 in 0.2 pH units. The color range was from yellow to blue, with shades of yellow-green and green-blue in between. A similar set of color tubes made from the actual culture media was prepared at CDC for other indicators as follows. The dehydrated culture medium (which included the acid-base indicator) was prepared at double strength (50 ml) and sterilized as recommended by the manufacturer, and 50 ml of 0.2M phosphate buffer[16] or 0.2M citrate buffer[16] (the amount of each ingredient of the buffer system varied according to the final pH desired) was added. This step was done while the medium was in a beaker (or a magnetic stirrer) with a pH electrode. Small adjustments to the final pH were made with 1N NaOH or HCl. For example, the malonate pH 7.2 color standard contained 50 ml of double strength malonate broth, 14 ml of 0.2M NaH_2PO_4 and 36 ml of 0.2M Na_2HPO_4. The pH was then adjusted to 7.2. Each of the color standards was then tubed in 100 x 13 mm borosilicate glass tubes (screw-cap), permanently sealed with a No. 000 rubber stopper and heated at 100C for five minutes to kill any vegetative bacteria. Contamination with spore formers was not a problem. Color standards were prepared for the following test media as described by Ewing and Davis:[7] mucate, malonate, lysine decarboxylase, methyl red, and carbohydrate fermentation medium with Andrade's indicator.

Automation of Biotyping

Biotyping can be done in conventional test tubes, in commercially available test systems,[23] or in several types of commercially available dishes. Figure 7-2 shows some of the dishes

---

*LaMott Chemical Products Co., Chestertown, Maryland.

Figure 7-1. A set of color standards (available commercially) for acid-base indicators.

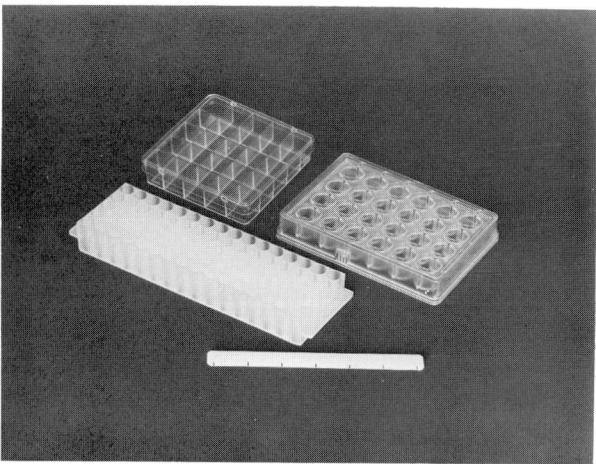

Figure 7-2. Three commercially available multiwell dishes, (clockwise from top left) A — the Dyos 25-well dish. B — The Linbro 24-well dish. C — The Gilson 80-well "fraction collector."

which can be adapted for biotyping: a 24-well disposable dish* (furnished sterile) that holds 2.5 ml in each well; a 25-well plastic dish† (furnished sterile); and an 80-well fraction collector made of (autoclavable) polypropylene.‡

Figure 7-3 shows a 61-well (each well holds 2 ml) dish which was made by drilling holes (10 mm diameter) into 13.3 cm blocks (3.8 cm thick) of autoclavable polycarbonate plastic.[5] The same dish was also used to hold 10 x 75 mm borosilicate glass test tubes (Fig. 7-3). Almost any biochemical test can be done in any of these dishes.

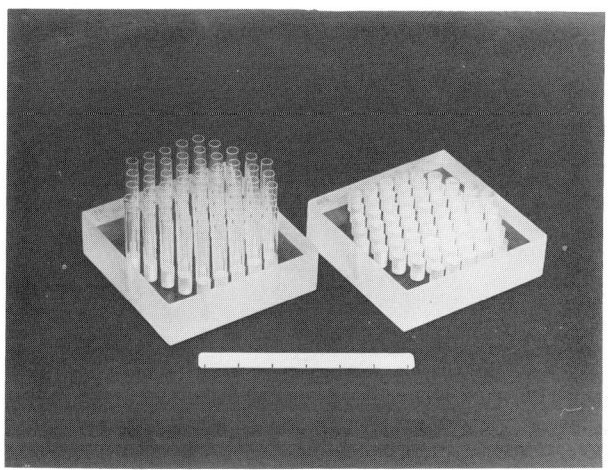

Figure 7-3. A 61-well plate which can be filled with agar or liquid media. The same dish containing 10- x 75-mm test tubes. In the first dish the biochemical tests are done directly in the plastic wells, but in the second dish the tests are done in the test tubes where there is less danger of spillage or cross contamination.

---
*From Linbro Chemical Co., Hamden, Connecticut.
†From Dyos Plastics, Surbiton, Surrey, England.
‡From Gilson Medical Electronic, Middleton, Wisconsin.
§AAA Plastics, Atlanta, Georgia.
§§From Johnny Brown Machine Shop, 4 Gold Star Acres, Northport, Alabama, 35476.

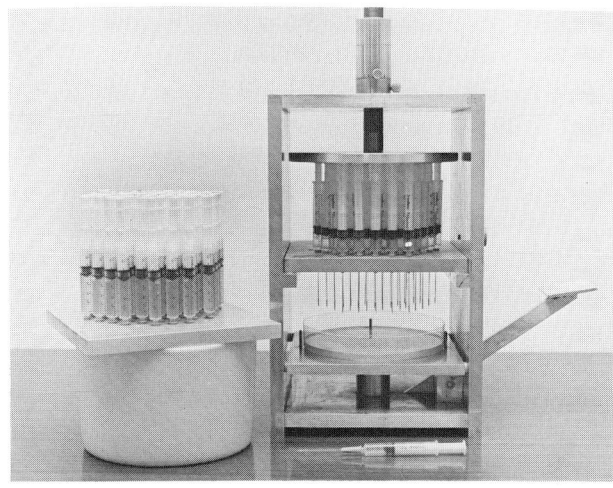

Figure 7-4. Multisyringe applicator for inoculating cultures into biotyping plates.

Figure 7-5. A template which holds 150 syringes (one syringe is in place) for multiple inoculation.

The 61-place inoculator[55] shown in Figure 7-4 can reduce the time required in inoculating the cultures into the dishes by about 90 percent. Bacterial suspensions are dispersed into the syringes, and a single drop (0.01 ml) is delivered from each syringe. The aluminum plate that holds the syringes is drilled to coincide with the centers of the wells on the plates. The cultures can also be inoculated (but the inoculum size is unknown) by pushing the needles into bacterial growth on a Petri dish and then lowering them into the test wells. About 150 cultures can be inoculated simultaneously onto a 150 x 15 mm Petri dish with the holder shown in Figure 7-5.

Results

Biotyping Nomenclature

Two different methods have been used for reporting biotyping results, and each has gained wide acceptance. The first system for defining biovars lists the name of the species followed by abbreviations for the biovar — for example, *Pseudomonas aeruginosa* mot⁻, nit⁻ would mean an isolate of *Pseudomonas aeruginosa* which is non-motile and does not reduce nitrate.

Many bacterial species are very "tight" in their biochemical reactions. For example, when forty-seven different biochemical tests were done on *Edwardsiella tarda*,[6] most of the strains had identical patterns. Of the forty-seven reactions, eleven were positive for 98 percent or more of the strains; thirty-four reactions were more than 99 percent negative. Only two reactions were variable; arabinose fermentation was 9.4 percent positive, and glycerol fermentation was 35.8 percent positive. Thus, a species which is "tight biochemically" is composed of strains which vary very little in their biovars. Almost all strains of *Edwardsiella tarda*, therefore, are H_2S^+, urea⁻, indole⁺, MR^+, VP^-, citrate⁻, etc., for the forty-seven reactions. A single strain which has the same biochemical reactions as most of the other strains in the species is called a "wild type" strain (wt). This would be written as *Edwardsiella tarda* wt. Another strain may vary slightly from the expected pattern: for example, an indole-negative strain could be considered a "wild type strain which does not produce indole in the standard test." This strain would be written *Edwardsiella tarda* ind⁻. Similarly, combinations can be formed — *Edwardsiella tarda* ind⁻ sorb⁺ malt⁻ (indole negative, sorbitol positive, maltose negative). This type of nomenclature is particularly useful for designating biovars in the "biochemically tight" species.

A different kind of nomenclature has proven more useful in designating biotypes in species which are not "biochemically tight." *Escherichia coli* is an example of a "biochemically variable" species. When several hundred strains of *E. coli* are tested biochemically, there is considerable variability among the individual strains. For example, the percent positive for some of the tests is: lysine 88 percent; arginine 17 percent; motility 70 percent; dulcitol 38 percent; and raffinose 54 percent. Only eighteen of forty-two biochemical tests are 90 percent or greater positive or negative for the strains; the rest are eleven — 89 percent positive. This is in contrast to *Edwardsiella tarda*, where the same reactions are almost always 99 percent positive or 99 percent negative. For *E. coli*, a true "wild type" cannot be defined because there is so much biochemical variation. In variable species like *E. coli*, it is better to define biovars in terms of the individual reactions, as illustrated in Table 7-I. A precise way to write the biovar of strain 1 for the nine biochemical reactions in the table would be *E. coli* biovar$^{(7\ day)}$100 030 016 (each number indicates the exact day that each reaction became positive). In a clinical laboratory where the reactions may be read at 24 h only, the same strain would be *E. coli* biovar$^{(24h)}$ 100 000 010. The delayed reactions for test 5 and 9 are not recorded because the culture was discarded after twenty-four hours.

There are a number of shorthand methods for changing the long pattern of biochemical results into shorter biovars (1,5,10). Thus, the data in Table 7-I can be reduced from nine reactions to three digit numbers by the shorthand given at the bottom of

Table 7-I. DEFINING BIOVARS FOR *E. COLI*

E. coli strain	\multicolumn{9}{c}{Results for Test:}									Shorthand* for:	
	1	2	3	4	5	6	7	8	9	24 h Biovar	7 d Biovar
1	+*	–	–	–	+³	–	–	+	+⁶	586†	564
2	–	–	+³	–	–	–	–	–	–	888	788
3	+	+	+	–	+	+	–	–	+⁷	148	147
4	+	+	+	–	+	+	–	+	–	146	146
5	+	+	+	–	+	+	–	+	–	146	146

*Symbols: + = positive within 24 hours. The day the reaction became positive is given by the superscripts of the +'s; "–" means the reaction was still negative on the last day it was determined (7 days in this case, but the last day of reading can be set at any value, even 24 hours).
†Shorthand: +++ = 1, ++– = 2, +–+ = 3, –++ = 4, +–– = 5, –+– = 6, ––+ = 7, ––– = 8; as proposed by Farmer.[10]

the table. In any shorthand, it is extremely important to define exactly what "+" and "–" mean. Table 7-I illustrates that Strain 1 is (in shorthand) "biovar 586" at twenty-four hours but "biovar 564" at seven days, because the seven-day biovar includes the two delayed reactions. In contrast, strain 4 is "biovar 146" at both 24 h and 7 d. Several different shorthands are used today to reduce data.[1,5,10,25] In Table 7-I, +++ = 1; in reference 1, +++ = 7, and in reference 25, +++ = 5. Thus, one must be careful in deciphering these different shorthands. The actual biochemical results for a given strain would be identical when the different shorthands are changed back to "+" and "–."

The Problem of "Fluctuating Biotypes"

Table 7-II shows that there is usually disagreement between CDC's results and those obtained by the laboratory which referred the culture. The ten cultures in Table 7-II were not chosen to show disagreement; they were ten consecutive cultures received in March 1976. There was disagreement on all ten, which means that the CDC biovar would be different from the sender's biovar in each instance. These differences are probably due to variations in media and methods rather than genetic change in the organism.

Table 7-II. DIFFERENCES IN TEST RESULTS BETWEEN CDC AND THE LABORATORY THAT SENT THE CULTURE

Culture	Number of Tests Compared	Number of Results That Disagreed	Percent Disagreement
1	20	4	20
2	26	4	15
3	30	1	3
4	32	2	6
5	19	2	11
6	20	2	10
7	13	1	8
8	32	3	9
9	32	9	28
10	25	2	8
		Average Disagreement	12

Temperature

In the past, many laboratories have used 37C as the standard temperature. However, many clinical laboratories have lowered this to 35C to enhance growth of species such as *Neisseria*. For us, as a reference laboratory, 36C is a compromise, thus we can say that our results are done at 36C ± 1C and we will be within 1 degree of those who incubate at 35C and those who incubate at 37C. There is little data comparing biotyping at 35C with that done at 37C, and the difference may or may not be significant. The usual laboratory thermometer can be ±2 degrees at 36C. All thermometers should be checked against one calibrated by the National Bureau of Standards. Otherwise, some incubators "set at 37C" could actually be set at 39C and some "set at 35C" could actually be set at 33C.

Influence of Media on Biotyping Results

Ideally, all media for biotyping should be chemically defined, prepared with ultra-pure water, and dispensed into completely inert containers. Unfortunately, the current state of the art is far from this. Most media for biotyping contain peptones, infusions, meat extracts, or yeast autolysates. Each of these is poorly defined chemically and can vary from batch to batch. One way to minimize these variables is to use the same lot number of each medium (purchase 10-50 lbs. of the same lot number, and use it throughout); however, this does not seem practical for the average laboratory. Another important variable in biotyping is "agar-agar." The amount of divalent cations and organic matter varies considerably among different brands of agar and even among different lot numbers of the same brand.[12] Whenever possible, a test should be done in a medium without agar, and the agar in many tests could easily be eliminated. For example, the test for utilization of citrate as the sole source of carbon is best done in a liquid rather than a solid medium. Much of the lot-to-lot variation in agar-containing media is probably due to the agar-agar.

The quality of water used in media for biotyping can also be a variable. Phycologists have known for many years that in the spring, algae bloom in lakes that are used as water supplies for many cities. Metabolic products from the bloom contaminate the water, and the metabolic products are not removed by deionization and often not even by distillation. A citrate test done in this enriched water is not really the same as one done in the fall after the products have been naturally degraded. Even if all other test ingredients have been standardized, water can be a variable and can cause different results. *Pseudomonas aeruginosa* can grow in even the highest quality of distilled water.[14] What chemicals is it using as sources of carbon and energy? Where does it get phosphorus, nitrogen, and sulfur for biosynthesis? No one has given a good answer to these questions, but we know that these must be in the water. Other laboratories use deionized water, which has inorganic but not organic constituents removed. In England, many laboratories even use tap water for media making.[4] Differences in distilled water from one laboratory to another are probably minor,[17] but the difference between tap water and glass-distilled water can cause considerable variation in the final test results.

Human Errors

Human mistakes can also cause errors in biotyping, and Table 7-III lists some of these. Often a "false positive" will cause suspicion, particularly if the "false positive" is a rare biovar — for example, H_2S^+ *Shigella sonnei*. This error can sometimes be discovered only after a long tedious investigation.[13] Table 7-III shows an incorrect result for *Serratia marcescens* on L-arabinose fermentation. The arabinose+ *S. marcescens* is suspicious because most strains are arabinose-. The reverse of this situation would be much harder to detect. A strain of arabinose- *S. liquefaciens* inoculated into the tube would probably have gone undetected, because this species is usually arabinose+. Thus an arabinose- biotype would be scored as arabinose+ because of the contaminant. This type of error is almost impossible to detect unless tests are repeated at least once.

Influence of Time or Inoculum

The time required for a reaction to become positive depends on the inoculum size and incubation time. For reproducibility,

Table 7-III. EXAMPLES OF INCORRECT BIOTYPING RESULTS AND REASONS FOR THEM

Incorrect Biovar	Explanation
Escherichia coli — Urea$^+$	Bad batch of medium, too sensitive
Salmonella typhi — H_2S^-	Done in 13 x 100 mm tubes rather than 16 x 125 mm tubes which are more sensitive for H_2S detection on TSI
Shigella sonnei — H_2S^+	Contaminated with an anaerobe
Enterobacter cloacae — D-Sorbitol$^-$	Sorbitol omitted from the fermentation tube
Salmonella typhimurium — Non-Motile	Motility medium not properly stirred before dispensing, agar concentration too high in this tube
Serratia marcescens — L-Arabinose$^+$	Tube contaminated with an organism which fermented arabinose
Shigella flexneri — Lysine$^+$	Not enough mineral oil for overlay
Enterobacter hafniae — Oxidase$^+$	Old culture used for test
Klebsiella pneumoniae — Citrate$^-$	The citrate medium was dispensed just after a medium with the antibiotic colistin had been dispensed. Antibiotic leached from tubing and inhibited the growth.
Proteus mirabilis — H_2S^-	H_2S^+ result was mistakenly recorded as H_2S^-
Yersinia enterocolitica — motile at 36C	Temperature in the incubator was actually 33C when measured with an accurate thermometer

both of these factors should be controlled, but an organism's biovar can be expressed at any time interval. In clinical laboratories, it is often convenient to take the final reading after 18-

20 h of incubation. This approach will usually lead to correct identifications, but the biovar will often be different from the biovar read at 24 and 48 h. Readings at 48 h will be more consistent than those at 18 h, because many reactions become positive between 18 and 48 h. Biovars at 48 h rather than 24 h would also be better epidemiological markers for the study of hospital-acquired infections. Table 7-I illustrates how time of incubation influences the biovar. These data show that a strain does not have an absolute or fixed biovar. Instead, the biovar must be defined in terms of the actual test conditions, and often the biovars are different at 18, 24, 48, and 168 h. This can be expressed as "biovar^{24h}" or "biovar^{7d}" which is the biovar at twenty-four hours and seven days respectively.

Influence of Test Procedure

Everyone has an idea of what a positive methyl red test is. However, descriptions of a "positive methyl red test" vary considerably. Table 7-IV gives six definitions by six different authors and shows that each definition is slightly different. None of these is an operational definition. An operational definition is one that is based on a measurable quantity. An example of an operational definition for a methyl red test would be "a pH of 5.70 or less, measured by a pH meter," which would eliminate the jargon about "cherry red" and "red-orange." Most authors do not give enough details in describing tests. For example, Table 7-V gives one author's description of the test for malonate utilization. Below this is a more detailed description which does not leave as much room for the reader to make incorrect decisions. Terms such as "slightly turbid," "red-orange color," "growth," or "alkaline" can easily be replaced with operational definitions such as "optical density" or "pH." It is not even necessary to measure these quantities in an instrument, because turbidity or pH color standards can be prepared and used in defining a positive test (Table 7-V).

Subjectivity in Calling a Test Positive or Negative

In biotyping, end-point determinations must be standardized

Standardization and Automation in Biotyping

Table 7-IV. DEFINITIONS OF A POSITIVE METHYL RED TEST BY SIX DIFFERENT AUTHORS

Author(s)	Definition
1	"Positive tests are bright red, weakly positive tests are red-orange, and negative tests are yellow or orange."
2	"Positive reaction is signified by a red colour."
3	"Red colour = +, *orange colour* = $\pm$, yellow colour = –."
4	"A red color is positive and a yellow is negative."
5	"Red colour = positive, yellow colour = negative"
6	"Positive . . . a distinct red, yellow color . . . is regarded as a negative reaction, while intermediate shades should be considered doubtful."

Table 7-V. TWO DESCRIPTIONS FOR DOING THE TEST FOR MALONATE UTILIZATION

Present Description

Medium:	Malonate Broth (Volume? Container?)
Inoculation:	"Inoculate from a young agar slant or broth culture."
Incubation:	37C; discard after 2 days
Reading:	"Positive results are indicated by a change in the color of the indicator from green to Prussian blue"

Description Which is More Operational

Medium:	Commercial Malonate Broth, 1.5 ml in 100 x 13 mm boro-silicate glass tubes; autoclave 121C, 15 min.
Inoculation:	10^4 cells, which are < 24 h old and have been grown on trypticase soy agar.
Incubation:	36 ± 1 C; read at 24 h and 48 h
Reading:	Compare with the set of pH color standards for the malonate test and score: + = pH of 7.2 or > – = pH < 7.2

for all tests in the schema. For example, author 1 in Table 7-IV gives the following instructions in interpreting the methyl red test: "Positive tests are bright red, weakly positive tests are red-orange, and negative tests are yellow or orange." This has been the standard definition used in the enteric laboratories at CDC for many years. To determine how this nonoperational definition of a positive test would be interpreted by different workers in our laboratory, I read this definition to fourteen different people who were working in the Enteric Section at CDC. Each person was then shown a series of color standards in 0.1 pH units for methyl red which had been adjusted from pH 7.0 (yellow) or 5.0 (bright red). These represented the gradual color change of methyl red indicator from yellow to bright red. Table 7-VI shows that there was a considerable difference of opinion as to what was a positive methyl red test.

Similarly, author 1 defines a positive malonate test as follows: "Positive results are indicated by a change in the color of the indicator from green to Prussian blue." This definition was read to the same group of workers, and they were shown color standards for malonate from pH 6.6 (green) to 8.0 (bright blue). From pH 6.9 to 7.4 the bromthymol blue indicator in the malonate medium gradually becomes blue through shades of blue-green. Table 7-VI shows the variation in this end point. The importance of these end-point differences in biotyping can be illustrated by the following example. A hypothetical culture is inoculated into the methyl red and malonate broths. After forty-eight hours when the two final readings are made, the broths have come to pH 5.8 ("red-orange") and 7.0 ("bluish-green") respectively. Worker number 3 would read this culture's biovar as "MR⁻, malonate⁻," but worker number 7 would call it "MR⁺, malonate⁺." The biovar of the culture did not really change; the discrepancy occurred because the two workers defined the end points differently. If the definition were changed to operational terms this type of discrepancy would be greatly reduced or eliminated. Two definitions that could be suitable substitutes for the nonoperational ones for methyl red would be:

Operational Definition 1: After incubation for forty-eight hours at 36C, the tube is removed from the incubator and a pH electrode (the pH meter should be standardized in critical

Table 7-VI. DIFFERENCES IN END-POINT DETERMINATION BY 14 LAB WORKERS IN ENTERIC BACTERIOLOGY

Test	pH for End Point	Number Reading pH as the End Point
Methyl Red	6.1 or >	0
	6.0	1
	5.9	1
	5.8	1
	5.7	3
	5.6	5
	5.5	2
	5.4	1
	5.3 or <	0
Malonate	6.8 or <	0
	6.9	1
	7.0	3
	7.1	2
	7.2	6
	7.3	2
	7.4 or >	0

range, at pH 5.6) is inserted. A pH of 5.60 to 5.70 is recorded as "+weak," a pH of 5.60 or less is recorded as "+," and a pH greater than 5.70 is scored as "-."

Although this definition is operational, it is not practical because of the time needed to make individual measurements with a pH meter. A more practical and only slightly less standard definition would be:

Operational Definition 2: After incubation for forty-eight hours, 0.2 ml of the (standard) methyl red indicator is added. The tube is shaken, and the color is compared to the ten standard color tubes for the methyl red test. A pH of 5.6 to 5.7 (based on visual comparison with these color tubes) is recorded as "+wk," a pH of 5.6 or less is recorded as "+" and a pH > 5.7 is called "-."

Almost all of the tests in biotyping could be defined in a similar fashion.

Variability in Each Person's Recognizing Their Own End Point

Table 7-VI showed each person's end point for the methyl red test. I then determined how well each person could recognize the same end point when shown a negative methyl red color standard only 0.1 pH unit higher than his or her end point. Table 7-VII shows that two workers could do this very consistently, but others varied in their ability to consistently recognize the end point they had chosen. Everyone, however, could consistently recognize tubes which were 0.2 pH units from their end points.

Automation in Biotyping

Automation by Surface Replication

One way that biochemical testing can be automated is by multiple inoculations of the bacterial cultures onto dishes filled with agar media.[2,15] In this approach, liquid test media are

Table 7-VII. INDIVIDUAL DIFFERENCES IN DETERMINING THE END POINT FOR THE METHYL RED REACTION

Person	Consistency in Recognizing Correct End Point*
A	100%
B	100
C	95
D	90
E	90
F	90
G	90
H	65
I	50

*Each person chose the end point. The person was then tested with this end point tube and a tube 0.1 pH unit higher to see how consistently the two could be differentiated.

completely replaced by solid media. This technique has two main problems. One is that diffusible metabolic products (acid, NH_3, H_2S, etc.) spread from the original point of inoculation and obscure reactions of adjacent cultures. The second is that swarming or spreading strains can spread across the entire plate, making the results impossible to interpret. Although these difficulties can be reduced somewhat by changing the media,[2,15] consistent biotyping results may be difficult. Biotyping in individual wells or tubes almost entirely eliminates these problems.

Automation of Biotyping in Individual Compartments

Figure 7-2 shows some dishes which are commercially available and can replace conventional test tubes with media with certain precautions (for example, urea agar should be overlayed with mineral oil, because when ammonia is released during urea hydrolysis, it diffuses to adjacent wells and causes a false positive test). Dishes 1 and 2 are made of polystyrene plastic and cannot be autoclaved, so sterile media is added to the already sterile dishes. Dish 3 was designed for fraction collecting, not for doing biotyping. It is made of polypropylene, which is autoclavable, but no top is commercially available. It is filled with media, placed in an autoclavable polypropylene bag* and autoclaved. A 2 ml Cornwall syringe is very useful for filling the individual wells of all these dishes.

A second method, which eliminates individual filling of each well, is called the "Mini Test Dish," which is illustrated in Figure 7-6. An agar medium is made, autoclaved, then cooled to 50C. It is poured into the bottom dish, and the sterile insert divider is added while the agar is still molten. After the agar hardens, thirty-six individual test areas are available for inoculation.

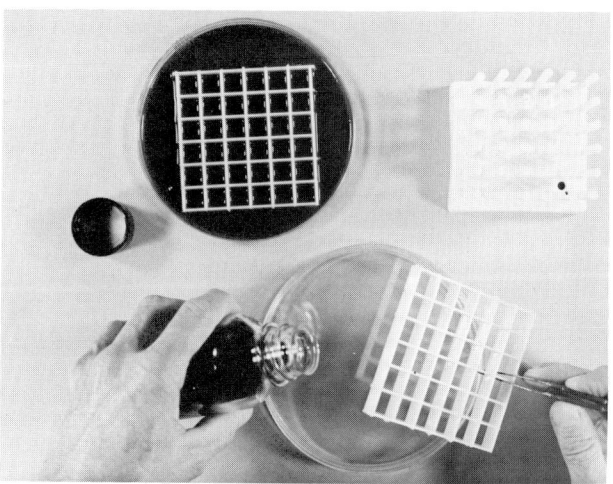

Figure 7-6. Two "Mini Test Dishes" — One has been prepared; the "insert divider" is being added to the other.

---

*Five sizes are available from Interex, 3 Strathmore Road, Natick, Massachusetts, 01760.

A slightly different approach to biotyping utilizes a holder which tightly holds glass or plastic tubes. This holder is illustrated in Figure 7-3. Blocks (3.8 cm thick) of autoclavable polycarbonate plastic were cut and drilled to hold 10 x 75 mm borosilicate glass tubes. Biochemical tests are done in tubes held in the plate and are thus identical to the standard "macro tests." Variables encountered in the plastic plates are eliminated. Figure 7-7 shows two other holders for automated biotyping.

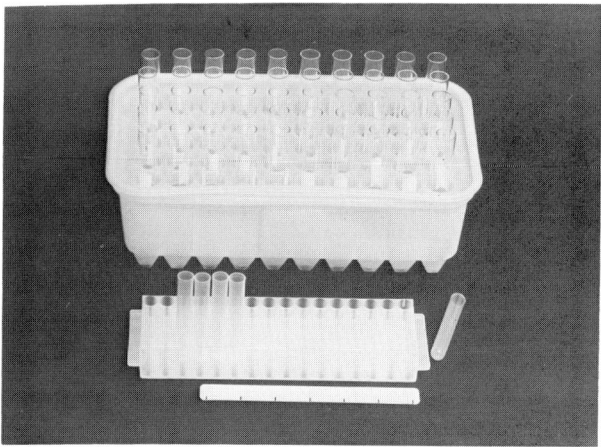

Figure 7-7. Two test tube holders which can be used in applying automation to biotyping. The one at the top is from Corning Glass Works, Corning, N.Y.; the one at the bottom is the same one shown in Figure 7-2.

Multiple Inoculation

The Steers replicator[31] has been useful in biotyping.[2,15] Figure 7-6 shows a "poor man's Steers replicator" which was made at a fraction of the cost of the Steers replicator (which is commercially available). A 102-peg test-tube rack* is sawed so that there are thirty-six pegs in each section (Fig. 7-8). The

*From Endicott-Seymour, Ann Arbor, Michigan.

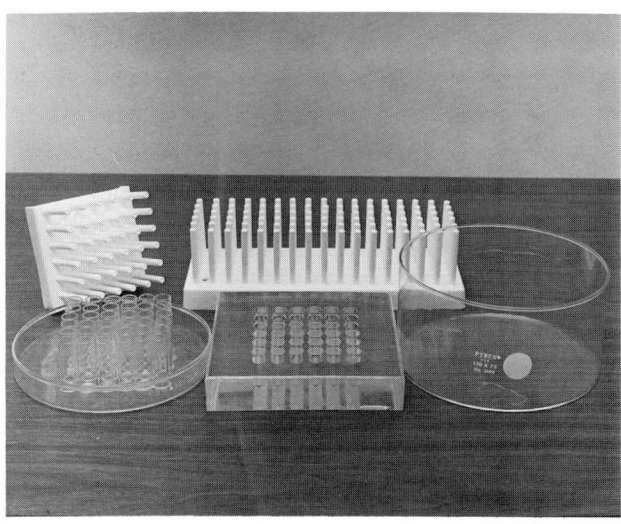

Figure 7-8. A 104-peg test tube rack, a 36-peg replicator which has been cut from it, two types of 36-well trays to hold cultures, and a glass dish which is filled with boiling water to decontaminate the replicator.

wells which hold the culture are made from a 2.5 cm thick polycarbonate plastic plate which is drilled with thirty-six holders whose centers coincide with the centers of the pegs. Up to thirty-six bacterial cultures are added to the wells, and they are inoculated onto plates or into wells with the thirty-six-peg inoculator. This inoculator replaces the more complicated and more expensive Steers replicator.

Another type of multi-inoculator is illustrated in Figure 7-3. This applicator contains autoclavable syringes which are filled with bacterial cultures and locked into place. A screw mechanism is turned which exerts pressure on the syringes, and a small drop of liquid is expressed from each one. A whole drop can be expressed and allowed to fall by gravity, or a partial drop can be "touched-off" (to an agar surface or well with liquid media) by raising the lower plate. Other accessories with this multipurpose applicator include a capillary inoculator (Fig. 7-9), similar to the one described by Hartman,[18] and a multi-needle inoculator (Fig. 7-10). These fit in the applicator (Fig. 7-4) in the same position as the syringe-plate.

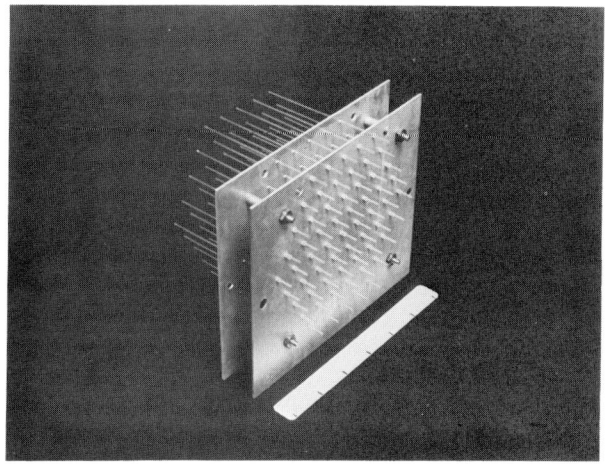

Figure 7-9. Capillary inoculator which is filled with cultures, inserted into the applicator shown in Figure 3, then used to inoculate plates.

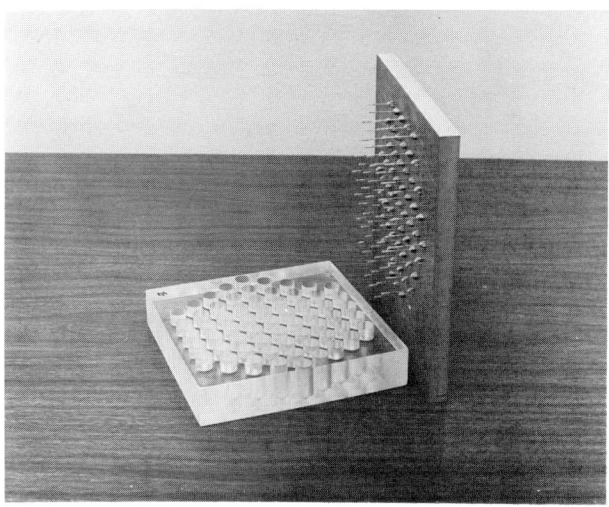

Figure 7-10. Multineedle inoculator. The 61 needles are touched to growth on a plate containing 61 different cultures, then used to inoculate the test plate shown in Figure 7-3.

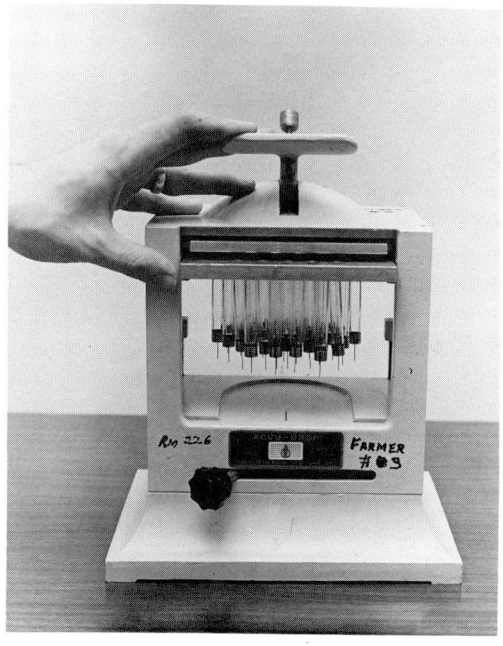

Figure 7-11. Accu-Drop applicator which contains 61 cartridges. The original applicator has been modified to hold 61 rather than 24 cartridges.

An additional applicator, which is commercially available,† is shown in Figure 7-11. The sixty-one cartridges are filled with different bacteria and then used to inoculate biochemical test plates. The pressure handle is depressed, and a single drop is delivered from each cartridge.[11]

Problems with Automation

Most of the problems in automated biotyping procedures result from the fact that the dishes used were not designed specifically for biotyping. A solution to this problem would be for a commercial company to make a dish with wells or tubes designed just for biochemical testing. A "biotyping cube" designed at CDC is an example of this approach (Fig. 7-12). This

---
†From Sylvania Co., Millburn, New Jersey.

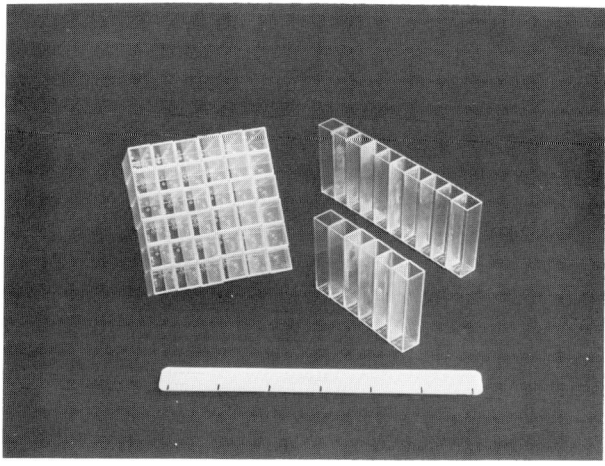

Figure 7-12. "Biotyping Cube." These are examples of biotyping plates that can be "injection moulded" out of plastic.

or similar dishes could then be made and used in automated biotyping with no loss in accuracy. Studies in many laboratories have shown that automation is feasible for biotyping, routine identification, and antimicrobial sensitivity testing.[2,12,17,18,27]

Discussion

I have pointed out some of the variables in biotyping and have suggested how careful standardization of the biochemical tests can eliminate them. In his recent paper on test reproducibility, Sneath[29] points out that intralaboratory variation in biochemical testing can be as high as 12 percent but is usually less than 4 percent. However, variation between laboratories was as high as 20 percent with a mean of about 10 percent. These figures are in agreement with our own experiences.

Many factors cause these discrepancies — media, distilled water, incubator temperatures, mislabeling, misrecording, differences in test procedures, end-point interpretation, and other errors, both human and nonhuman. Unfortunately, it is difficult to detect many of these errors. Until some of the variables are corrected, biotyping will remain a poorly standardized technique.

Since biotyping is now being used as a primary method of "fingerprinting" in hospital epidemiology, a few comments about variability are in order here. A typical situation could be summarized as follows:

> There are four patients in the intensive care unit. Three have blood cultures positive for *Klebsiella pneumoniae*. The biovars (at 24 h) are, respectively: Isolate 1 — 631 561 (18 reactions changed to 6-digit code), Isolate 2 — 631 561 and Isolate 3 — 631 562. When *K. pneumoniae* biovar 631 561 was checked with previous records, it was found to be very rare — only two isolates among 320 previous ones were this biovar. Thus, the probability of two patients having this rare biovar by chance alone is very small (less than 1 in 10,000). Cross-infection from one patient to the other could have occurred, or perhaps a common source (such as contaminated equipment or food) introduced the strain. Because 631 561 is a rare biovar in this hospital, it can be easily followed. If biovar 631 561 were common (say 50% of the isolates of *K. pneumoniae* in this hospital), then the chance of two patients acquiring it by chance alone would be 1 in 4. This latter biovar would be too common to be a good "epidemiological marker." Was the third patient (with *K. pneumoniae* biovar 631 562) related to this small outbreak? Biovar 631 562 differs from 631 561 by only 1 reaction out of eighteen. It may really be the same strain as the one from the other two patients, but it may not be. The antibiograms should be compared, which may answer the question, but a more definitive answer would come from serotyping.

This example illustrates a very important principle in biotyping: it is much more significant to have organisms with the same biovar (particularly if it is an uncommon one) than it is to have organisms with different biovars. A corollary to this principle would be — two isolates can be the same strain (in a genetic and epidemiological sense), even if they are different biovars. This is because of all the possible variables in biotyping. The results become more significant, however, if the biotyping results are reproducible when repeated under carefully controlled conditions.

Sneath also discusses test reproducibility in relation to correct identification.[29] Reproducibility is much more important in biotyping, because even a single incorrect result changes an organism's biovar. If fifty tests are done in the biotyping, two tests will be incorrect in most laboratories (assuming Sneath's 4% error rate). More and more biotyping is now being done with systems from commercial companies.[23] It will be interesting to see if these manufacturers can reduce some of the variables in biotyping through careful media selection, internal quality control, and standardized end point determination.

I have presented some thoughts on how biotyping can be automated. With conventional test-tube media, it takes about two hours to set up, inoculate, read (each day for 7 days), and record fifty biochemical tests on a single culture. When many cultures are processed at the same time, errors begin to occur because biotyping is a tedious process. Automation can both reduce costs and increase accuracy. As the demand for biotyping increases, commercial companies will develop ingenious solutions to problems encountered in biotyping. Conceivably, a culture could be injected into an instrument, and within four to twenty-four hours a complete identification, antibiogram, serovar, phagovar ("phage type"), and biovar could be printed. When this point is reached the applications of biotyping may be unlimited.

REFERENCES

1. API. *Quick Index for Common Profiles of Enterobacteriaceae.* Analytab Products Inc., Plainview, N Y, 1976.
2. Chadwick, P., G. J. DeLisle and M. Byer. Biochemical identification of hospital enterobacteria by replica agar plating. *Can J Microbiol, 20*:1653-1664, 1974.
3. Chester, B. and G. Stotzky. Temperature-dependent cultural and biochemical characteristics of rhamnose-positive Yersinia enterocolitica. *J Clin Microbiol, 3*:119-127, 1976.
4. Cowan, S. T. and K. J. Steel. *Manual for the Identification of Medical Bacteria, 2nd Ed.* New York, Cambridge Univ. Press., 1975.
5. Dito, W. R., J. Bulmash, J. Campbell and E. Roberts. *A numerical coding and identification system for the Enterobacteriaceae.* Chicago,

American Society of Clinical Pathologists, 1972.
6. Edwards, P. R. and W. H. Ewing. *Identification of Enterobacteriaceae. 3rd Ed.* Minneapolis, Burgess Publ. Co., 1972.
7. Ewing, W. H. and B. R. Davis. Media and tests for differentiation of Enterobacteriaceae. Atlanta, Center for Disease Control, 1970.
8. Ewing, W. H., A. C. McWhorter, G. A. Huntley and G. J. Hermann. A lysine negative strain of *Salmonella typhi*. *Publ Hlth Lab*, *30*:98-99, 1972.
9. Falcao, D. P., L. R. Trabulsi, F. W. Hickman and J. J. Farmer, III. Unusual Enterobacteriaceae: Lactose-positive *Salmonella typhimurium* which is endemic in São Paulo Brazil. *J Clin Microbiol*, *2*:349-353, 1975.
10. Farmer, J. J. III. Mnemonic for reporting bacteriocin and bacteriophage types. *Lancet*, *2*:96, 1970.
11. Farmer, J. J. Improved bacteriophage — bacteriocin applicator. *Appl Microbiol*, *20*:517-518, 1970.
12. Farmer, J. J. III, F. W. Hickman and J. V. Sikes. Automation of *Salmonella typhi* phage typing. *Lancet*, *2*:787-790, 1975.
13. Farmer, J. J. III, C. F. Riddle, M. D. Stargel, T. Iida, T. Aikawa, D. Achanzar and W. I. Taylor. Unusual Enterobacteriaceae: $H_2S^+Shigella$ *sonnei* — one authentic and one false positive due to contamination with the obligate anaerobe *Eubacterium lentum*. *J Clin Microbiol*, *3*:206-208, 1976.
14. Favero, M. S., L. A. Carson, W. W. Bond and N. J. Peterson. *Pseudomonas aeruginosa:* Growth in distilled water from hospitals. *Science*, *173*:836-838, 1971.
15. Fuchs, P. C. The replicator method for identifying and biotyping of common bacterial isolates. *Lab Med*, *6*:6-11, 1975.
16. Gomori, G. Preparations of buffers for use in enzyme studies. p. 138-146. In *Methods in Enzymology, Volume 1*. New York, Academic Press, 1955.
17. Hartman, Paul A. Miniaturized microbiological methods. *Supplement 1 to Advances in Applied Microbiology*. New York, Academic Press, 1968.
18. Hartman, P. A. and P. A. Pattee. Improved capillary-action replicating apparatus. *Appl Microbiol*, *16*:151-153, 1968.
19. Hickman, K., I. White, and E. Stark. A distilling system for purer water. *Science*, *180*:15-25, 1973.
20. Inolex Biomedical Division. *Inolex: Var-Ident System*. Glenwood, Illinois, Inolex Corp., 1974.
21. Kauffman, F. *The bacteriology of Enterobacteriaceae*. Baltimore, Williams & Wilkins, 1966.
22. Lapage, S. P., P. H. A. Sneath, E. F. Lessel, V. B. D. Sherman, H. P. R. Seeliger and W. A. Clark (Eds.). *International Code of Nomenclature of Bacteria*. Washington, D.C., American Society for Microbiology,

1976.
23. Nord, C. E., A. A. Lindberg and A. Dahlbäch. Evaluation of five test kits — API, Auxo Tab, Enterotube, Patho Tec and R/B — for identification of Enterobacteriaceae. *Med Microbiol Immunol, 159*:211-220, 1974.
24. Penner, J. L., N. A. Hinton and J. Hennessy. Biotypes of *Proteus rettgeri*. *J Clin Microbiol, 1*:136-142, 1975.
25. Rennie, R. P. and I. B. R. Duncan. Combined biochemical and serological typing of clinical isolates of *Klebsiella*. *Appl Microbiol, 28*:534-539, 1974.
26. Roche Diagnostics. *Enterobacteriaceae: Numerical Coding and Identification System for Enterotube*. Nutley, N. J., Roche Diagnostics, 1973.
27. Sielaff, B. H., E. A. Johnson and J. M. Matsen. Computer-assisted bacterial identification utilizing antimicrobial susceptibility profiles generated by Autobac 1. *J Clin Microbiol, 3*:105-109, 1976.
28. **Skerman, V. B. D.** *A Guide to the identification of the genera of bacteria.* Baltimore, Williams & Wilkins Co., 1967.
29. Sneath, P. H. A. Test reproducibility in relation to identification. *Int J Sys Bacteriol, 24*:508-523, 1974.
30. Society of American Bacteriologists. *Manual of Microbiological Methods.* N Y, McGraw-Hill Book Company, 1957.
31. Steers, E., E. L. Foltz, and B. S. Graves. An inocula replicating apparatus for routine testing of bacterial susceptibility to antibiotics. *Antibiot Chemother, 9*:307-309, 1959.
32. Wilson, G. S. and A. A. Miles. *Topley and Wilson's Principles of Bacteriology and Immunity.* 5th Ed. Baltimore, Williams & Wilkins Co., 1964.

Chapter 8

PERSPECTIVE OF BACTERIAL BIOTYPING FOR CLINICAL MICROBIOLOGY

Henry D. Isenberg and Albert Balows

MANY factors influence the need for and the usefulness of biotyping in the clinical microbiology laboratory. The most compelling of these is the desire to recognize that a particular bacterium is threatening the safety of patients, the medicolegal equanimity of an institution, or the health of others within the community. This laudable desire must be balanced, however, by the knowledge that formal classification that is easily applied to multicellular macroscopic biological systems cannot be applied equally to bacteria at the species and subspecies level. One must heed the caveats of Stanier and van Niel[14] concerning the absence of a clear concept of a bacterium and the restrictions imposed by the nature of prokaryotic cellular organization. In fact, within this century, the very use of the cell concept has been challenged by Dobell[3] who viewed the total of all vital and integrated expressions of any living organism as the basis for consideration as a separate entity and, thus, described the prokaryotic bacteria and blue green algae as "acellular" organisms.

The appreciation of the ability of prokaryotic organisms to carry on the functions associated with life within what is conventionally called a cell makes the considerations of Kluyver[10] and Lwoff[11] concerning the unity of biochemistry and biological order more meaningful. The Lwoffian definition accentuates the most basic similarities shared by all living forms: (1) a unity of plan represented by a nucleus embedded in protoplasm, (2) a unity of function based on similar essential metabolic pathways, and (3) a unity of composition which recognizes that the main macromolecules of all living forms are composed of the same small molecules. The so-called primitive nature of the prokaryotes, therefore, reflects their original ability to impose organizational order on prebiological organic molecules and to lay down the foundations which unify the diverse and complex expressions of animal and plant as well as protistal life.

But this unifying theme cannot dispel the simple fact that the individual bacterium is not the object of clinical laboratory analysis but rather the vast populations of bacteria we designate as colonies. It may well be that the assumption underlying the laboratory isolation of an organism, namely, that each separate colony constitutes the progency of a single bacterium, is correct. Nevertheless, the colony is composed of millions of cells, each reflecting the genetic potential of the particular isolate and able to express this capability in the appropriate environment. Since even rare mutations can be expressed in these large populations at any time, subtle differences between individuals within the colony may exist. Similarly, the proportion of different individuals in colonies, formed after transfer from a broth culture, may also be different for each such colony despite the seeming identity of the macroscopic expressions of microbial anatomy. This dilemma is complicated further by the potential influence of plasmid- or episomal-transmitted characters as well as the difference between the host and laboratory environment which might select different segments of the larger population identified by a species designation.[6]

DNA relatedness in such colonies may be the final proof which taxonomists require to establish the limits of a prokaryote species, but this type of analysis is well outside the present capability of most clinical laboratories. Furthermore, the sensitivity of this analysis has not progressed to the level where subspecies differences can be readily established. The scope and methodology of DNA analysis requires simplification and thorough study before it can be of use routinely in medical microbiology. This is especially true at this time when there is considerable political clamor for cost containment and microbiological analyses specifically directed to the care of the individual patient rather than the protection or benefit of other patients and hospital employees or of the community.[1,12] The impact of such external constraints does not portend too well for rapid resolution of problems requiring the application of molecular biology to medical and public health microbiology.

The problem of biotypes of bacteria is complicated still further by lack of agreement among a number of experts in microbial taxonomy. In enunciating "the fundamental principles of the realistic classification and nomenclature," Kauffmann[9] presents cogent arguments that identification based on morphological and biochemical analyses cannot establish a *real* species, but only biotypes or large groups of bacteria, such as subgenera, genera, etc. In his view, two biochemically identical cultures diagnosed in the orthodox classification manner as one species can belong to two different serotypes. In Kauffmann's classification these organisms represent two different species. The converse also obtains. Kauffmann would designate two cultures with the same serotype but different biochemical profiles as two biotypes of the same species. Kauffmann states that the realities of nature force the decision of a species, not the Tomistic or even Platonic reality decried already by John Locke[2] in his assertion: "Genera and species depend on such collections of ideas as man has made and not on the real nature of things."

Medical and public health microbiologists[2,7] have expressed the need for a realistic or pragmatic approach to bacterial taxonomy. The accepted authoritative works are based on hierarchical and biochemical principles difficult to apply to the everyday analyses performed in clinical microbiology laboratories.

Kauffmann bases his classification on the species as the fundamental unit, defined by serological, biochemical and other properties displayed by a group of cultures or types. The types which make up the groups do not belong to different taxa but represent properties of the same species, i.e. the ability to display various and sundry biochemical activities without requiring the establishment of subspecific or infrasubspecies categories. Kauffmann is adamant in his insistence that species cannot be determined without serological information. He states categorically that the species is the only category or taxon that occurs in nature and that this specified species is not an abstraction but a reality. Unfortunately, this concept is not explained in detail but is supported with a series of examples of great rational appeal to the microbiologist confronted with the task of identifying bacteria from pathological materials. Kauffmann dismisses all other attempts at classification as "creations of our minds" including such "vertical" or hierarchical systems as subgenus, genus, tribe, family, etc.

The essence of Kauffmann's approach is his call for a more pragmatic look at classification and nomenclature with the substitution of rational systems for the *International Code of Nomenclature of Bacteria*. He quotes the damaging admission of R. E. Buchanan and N. E. Gibbons in their introduction to the eighth edition of *Bergey's Manual of Determinative Bacteriology* that a meaningful hierarchy of bacteria is impossible. Although such a statement should temper the application of Bergey's guidelines, the practicing microbiologist continues to be hampered by the general noncritical acceptance of the classification and nomenclature the *Manual* proposes.[7]

Cowan[2] points out that Kauffmann's earlier views — which are the basis for his latest conclusions[9] — have not found universal favor with bacterial taxonomists, many of whom remain confused about the actual definition of the species in Kauffmann's terms, i.e. is a species a subdivision of a serotype or the smallest unit that can be characterized by present-day techniques? Huxley's[4] definition of a species could not pertain to bacterial and myxophycean species according to van Niel[15] who pointed out that the species ceases to be an actuality in agamic biological complexes. Although it is true that limited sexual exchange can be observed intra- and inter-specifically between certain bacteria, it also follows that plasmid-mediated expression of extrachromosomal DNA obfuscates potential conclusions and confirms the earlier observations of Shimwell and Carr[13] that the "existing classification based on the study of cultures under chemical and physical conditions may be nothing more than the classification of mixed and composite cultures which may vary in cell composition according to the tests applied."

The practical experience of public health and clinical microbiologists confirms this statement on several levels. In a collaborative evaluation of one of the earliest system approaches to the recognition of Enterobacteriaceae,[8] a very high percentage of agreement was found in test results with a commercial system. The disagreements encountered were concentrated in the supplemental battery of biochemical tests performed with "in-house" prepared substrates. Although the actual test substrate was identical, the milieu for the performance of the tests differed. Surprisingly, the results obtained differed repeatedly and led the authors to conclude that bacteria do not respond identically on ostensibly the same medium with only minor differences in formulation or with different indicators even if the major substrate, e.g. the name of the test medium, is the same. They emphasized the need for analysts to agree beforehand on the so-called conventional procedures against which the performance of an untested method is to be compared or to actually exchange media and employ identical materials so that then conclusions would be meaningful. The implications of these observations for clinical and public health microbiology laboratories are quite obvious.

Completely standard media or procedures which would permit uncritical comparison of identification between laboratories do not exist. The very material which enables the microbiology community to study microbial populations in that most unnatural condition, the pure culture, has not been defined in terms of chemical composition or physical condition. Today we are almost totally ignorant of the minimal growth requirements of most of the medically significant bacteria (if we can "domesticate" them at all). We are totally in the dark concerning optimal growth requirements of these organisms. Therefore, only the in-use evaluation of important laboratory media with laboratory-adapted reference strains remains as the sole manner of evaluating their suitability for use in the recognition of medically significant microorganisms.

Other aspects of this problem are not solved by standardized substrates or procedures which complicate the labor of the clinical and public health microbiologist. These considerations, previously outlined,[6] encompass the identity rather than the relatedness of bacteria cultivated in the laboratory with those active in the host. No one has tested bacteria in the natural environment for the same end products featured so prominently in laboratory identification steps. The question of the selection of laboratory variants from a larger population and pathogenic variants of a specific bacterium in the same host has not been resolved. We are not certain that the microorganism during the exponential growth phase in any culture is the same as the one capable of physiological function during the decline of the stationary phase in this same culture. Similarly, the selective role played by specific host organs in selecting an autochthonous microbiota as contrasted with the property of other organs to attract these microorganisms during a diseased state has not been resolved.

In addition, we must recognize[5] that our taxonomic skills are based primarily on observations with organisms isolated from pathological specimens. The limits of biochemical, physiological, immunological, and even bacteriophage-bacterial host behavior of microbial populations in their natural habitats have never been determined. Instead, the bulk of our information is derived from laboratory isolates which defy in some manner the existing artificial taxonomic groups and finally reach the attention of the experts in nomenclature in reference laboratories. Even the percentages of a given characteristic derived from testing large numbers of isolates which appear to diminish if not abolish the ritualistic adherence to dichotomous keys and which, more importantly, are being applied to computer technology as well as mathematical matrices to establish the most probable identity of an isolated culture are based largely on findings with bacteria selected by and isolated from disease conditions. There have been no serious attempts to determine the discriminatory capabilities of laboratory technology with natural and, therefore, mixed microbiota. Even pure cultures of bacteria separated from their natural sources have been studied only very rarely. Thus, results based on a consideration of variables used to determine characteristics employed in classification are not available. Instead, we have used the intruders into the intimate human biosphere, selected by favorable conditions in certain, often diseased, hosts to establish the criteria for identification.

Despite these strong arguments in support of a microbiological uncertainty principle, the practical measurement of mostly biochemical and physiological activities of bacteria in undefined mixtures of nutrients bolstered by select substrates has enabled practicing clinical microbiologists to call the same isolate by the same name repeatedly. Obviously, no rules of nomenclature or system of taxonomy can adequately explain why certain populations of prokaryotes behave similarly time and again to accommodate a previously designated genus and species. No one can expect the microbiologist to ignore this phenomenon, especially since the clinical application of such information is most helpful in the diagnosis of disease and the therapy of patients. One can only encourage attempts to delineate bacteria still further by using the same method, since additional information could become available. Such information may be helpful in the protection of other patients and hospital personnel, as well as the community at large. Such attempts may also provide useful information on the epidemiology and pathology of that particular group of microorganisms sharing an expanded set of characteristics. It may well be a Ptolemaic view of the microbial universe, but explanations for its functions are advanced and proved. The microbiologist must continue his significant efforts with methods and approaches for which no one has a totally satisfactory explanation.

REFERENCES

1. Bartlett, R. C. Medical microbiology: How fast to go — how far to go. *Significance of Medical Microbiology in the Care of Patients.* V. Lorian (Ed.). Baltimore, Williams & Wilkins Co., 1977, pp. 15-35.
2. Cowan, S. T. The microbial species — a macromyth. *Symp Soc Gen Microbiol, 12:*433-455, 1962.
3. Dobell, C. C. The principles of protistology. *Arch Protistente, 23:*269-

310, 1911.
4. Huxley, J. S. Introductory: towards the new systematics. *The New Systematics.* J. S. Huxley (Ed.). Oxford, Clarendon Press, 1940, pp. 1-7.
5. Isenberg, H. D. Biochemical rapid identification of Enterobacteriaceae. *Modern Methods in Medical Microbiology.* J. E. Price, J. Bartola and H. Friedman (Eds.). pp. 41-49. Baltimore, University Park Press, 1975.
6. Isenberg, H. D. Some cautionary considerations for the automation of clinical microbiologic procedures. *Mt. Sinai J Med, 44:*134-141, 1977.
7. Isenberg, H. D. and B. G. Painter. Indigenous and pathogenic microorganisms of man. In *Manual of Clinical Microbiology,* 2nd ed. E. H. Lennette, E. H. Spaulding and J. P. Truant (Eds.). Washington, D.C., American Society for Microbiology, 1974, pp. 45-58.
8. Isenberg, H. D., P. B. Smith, A. Balows, B. G. Painter, D. C. Rhoden and K. Tomfohrde. r/b Expanders: their use in identifying routinely and unusually reacting members of Enterobacteriaceae. *Appl Microbiol, 27:*575-583, 1974.
9. Kauffmann, F. *Classification of Bacteria.* Copenhagen, Munksgaard, 1975.
10. Kluyver, A. J. *The Chemical Activities of Microorganisms.* London, University of London Press, 1931.
11. Lwoff, A. V. *Biological Order.* Cambridge, The M.I.T. Press, 1962.
12. McCabe, W. R. Enterobacteriaceae: Clinical significance of speciation. *Significance of Medical Microbiology in the Care of Patients.* V. Lorian (Ed.). pp. 87-95. Baltimore, Williams & Wilkins Co., 1977.
13. Shimwell, J. L. and J. G. Carr. Are species of bacteria unclassifiable? *Leeuwenhoek ned Tydschr, 26:*383, 1960.
14. Stanier, R. Y. and C. B. van Niel. The concept of a bacterium. *Arch f Mikrobiol, 42:*17-35, 1962.
15. van Niel, C. B. Classification and taxonomy of the bacteria and blue-green algae. *A Century of Progress in the Natural Sciences — 1853-1953.* San Francisco, California Academy of Sciences, 1955.

AUTHOR INDEX

A

Aber, R. C., 44-45
Achanzar, D., 96
Agterberg, C., 24
Aikawa, T., 96
Ajello, L., 32
Allen, N., 44
Allen, S. D., 47
Allerhand, J., 24
Alonso, J. M., 24
Anditi, L. J., 67
Armfield, A. Y., 47
Atteberry, H. R., 66

B

Baker, C. J., 44
Ball, L. C., 45
Balows, Albert, 3, 66-67, 98, 105
Barnes, W., 45
Barrett, F. F., 44
Bartlett, R. D., 104
Bartola, J., 105
Bassett, D. C. J., 45
Baumann, P., 31
Beachey, E. H., 44
Bejot, J., 24
Bercovier, H., 24
Bergner-Rabinowitz, S., 45
Besdine, R. W., 44
Bobo, R. A., 31
Bodily, H. L., 66
Bond, W. W., 96
Bottone, E. J., 24
Breed, R. S., 66
Brenner, Don J., 12, 15, 24, 31
Brown, C., 24
Buchanan, R. E., 24, 31, 66, 101
Bulmash, J., 95
Byer, M., 95

C

Campbell, J., 95
Carr, J. G., 67, 101, 105
Carson, L. A., 96
Carter, R. F., Jr., 31
Cato, E. P., 66
Chadwick, P., 95
Chester, B., 24, 95
Chow, A. W., 66
Clark, W. A., 96
Cobbs, C. G., 67
Cowan, S. T., 12, 24, 70, 95, 101, 104
Cunningham, M. W., 44

D

Dahlbach, A., 97
Davies, A. M., 45
Davis, B. R., 70-71, 96
DeHaan, R. M., 66-67
DeLisle, G. J., 95
Dito, W. R., 95
Dobell, C. C., 98, 104
Doggett, R. G., 31
Doudoroff, M., 29, 31-32
Dowell, V. R., Jr., 47, 49, 53, 59, 61, 66-67
Duncan, I. B. R., 97

E

Edwards, L. R., 45
Edwards, P. R., 24, 70, 96
Egan, R. G., 46
Engh, G. A., 34, 45
Everly, W. N., 45
Ewing, W. H., 16, 18, 24, 70-71, 96

F

Facklam, R. R., 37, 44-46

Falcao, D. P., 24, 96
Falkow, S., 24
Fanning, G. R., 24
Farmer, J. J., 96
Farmer, J. J., III, 31-32, 68, 96
Farmer, L. H., 31
Favero, M. S., 96
Fayez, J., 45
Feeley, J. C., 45
Felner, J. M., 53, 59-60, 66
Fife, M. A., 16, 24
Finegold, S. M., 66-67
Finlayson, Margaret, 68 fnt.
Fischetti, V. A., 44
Flora, D. J., 66
Foltz, E. L., 97
Fox, E. N., 44
Franciosi, R. A., 44
Fraser, C. A. M., 45
Friedman, H., 105
Fuchs, P. C., 96

G

Gibbons, N. E., 24, 31, 66, 101
Gilardi, G. L., 32
Gomori, G., 96
Gordon, R. C., 44
Gotschlich, E. C., 44
Graves, B. S., 97
Guinee, P. A. M., 24
Guze, L. B., 66-67

H

Hall, R. T., 45
Hare, R., 45
Harris, D. J., 45
Harrison, G. M., 31
Hartman, Paul A., 90, 96
Hawkins, T. M., 49, 61, 66
Hayes, P. S., 45-46
Hejda, V., 46
Hennessy, J., 97
Hermann, G. J., 96
Hickman, F. W., 96
Hickman, K., 96
Hinton, N. A., 97
Hoadley, A. W., 32

Hoffman, N. J., 67
Holdeman, L. V., 52, 66
Howell, J. T., 44
Huntley, G. A., 96
Huxley, J. S., 101, 105

I

Iida, T., 96
Isenberg, Henry D., 3, 98, 105

J

Jelinkova, J., 46
Johnson, E. A., 97
Johnson, R. H., 45
Johnson, W. D., 67
Jones, L. F., 31-32
Jones, W. L., 46
Juni, E., 32

K

Kasper, D. L., 44
Kauffmann, F., 12, 70, 96, 100-101, 105
Kaye, D., 67
Kluyver, A. J., 98, 105
Knapp, W., 15, 24
Knittel, M. D., 24
Knostman, J. D., 44
Ko, H. L., 60, 67
Krishnan, L., 45
Kronus, R., 46

L

Laënnec, Rene, 3
Lancefield, Rebecca C., 33, 45
Lapage, S. P., 96
Lennette, E. H., 27, 32, 105
Lessel, E. F., 96
Levin, J., 67
Lindberg, A. A., 97
Locke, John, 100
Lombard, G. L., 47, 67
Lorian, V., 104-105
Lukesova, M., 46
Lwoff, A. V., 98, 105

M

MacLowry, James D., 6
Malowany, M. S., 24
Martin, W. J., 67
Mason, J. O., 66
Masterson, N., 32
Matsen, J. M., 97
Maxey, Joy, 47 fnt.
Maxted, W. R., 37, 45
McCabe, W. R., 105
McCarty, M., 45
McWhorter, A. C., 96
Miles, A. A., 97
Miller, G. L., 45
Mollaret, H. H., 24
Moody, M. D., 46
Moore, H. B., 67
Moore, W. E. C., 48, 52, 66
Murphy, K. E., 45
Murray, E. G. D., 66

N

Newton, E. J., 31
Niléhn, B., 14, 24
Nord, C. E., 97

O

Ofek, I., 45

P

Padula, J. F., 44
Painter, B. G., 105
Palleroni, N. J., 29, 32
Papageorgis, C., 32
Pattee, P. A., 96
Paule, R., 67
Penner, J. L., 97
Peterson, N. J., 96
Phillips, S. E., 67
Pine, L., 44
Price, J. E., 105
Puhvel, S. M., 67
Pulverer, G., 60, 67

Q

Quintiliani, R., 34, 45

R

Reiser, R. M., 67
Rennie, R. P., 97
Rhoden, D. C., 105
Rhodes, P. G., 45
Riddle, C. F., 96
Roberts, E., 95
Rosebury, T., 67
Russell, H., 45

S

Sconyers, B. J., 44
Seeliger, H. P. R., 96
Seidler, R., 19
Seidler, R. J., 24
Shaw, D. A., 67
Sherman, V. B. D., 96
Shimwell, J. L., 101, 105
Sielaff, B. H., 97
Sikes, J. V., 96
Simmers, R. W., 67
Siviglia, G., 44
Sjostrom, B., 14, 24
Skerman, F. J., 24
Skerman, V. B. D., 70, 97
Skinner, F. A., 67
Skjold, S. A., 45
Smith, L. D. S., 67
Smith, N. R., 66
Smith, P. B., 105
Sneath, P. H. A., 93, 95-97
Sonnenwirth, A. C., 67
Spaulding, E. H., 32, 105
Speltie, T. M., 24
Stanier, R. Y., 29, 31-32, 98, 105
Stargel, M. D., 47, 67, 96
Stark, E., 96
Starr, S. E., 67
Steel, K. J., 70, 95
Steere, A. C., 45
Steers, E., 97
Steigerwalt, A. G., 24
Stinnett, J. D., 32
Stotzky, G., 24, 95
Sutter, V. L., 66-67
Swartz, Stephen, 41 fnt.
Swenson, Jana, 47 fnt.

T

Taylor, W. I., 96
Thacker, L. G., 44
Thal, E., 15, 24
Thomas, E. T., 32
Thompson, F. S., 47, 67
Tomfohrde, K., 105
Topley, 70
Trabulsi, L. R., 96
Truant, J. P., 32, 105
Turkova, S., 46

U

Updyke, E. L., 66

V

Valchova, M., 46
van der Waaij, D., 24
van Niel, C. B., 98, 101, 105
Vosti, K. L., 45

W

Wannamaker, L. W., 45

Warford, L. R., 45
Washington, J. A., 67
Watson, E. D., 67
Wauters, G., 14, 25
Weaver, R. E., 24
White, I., 96
Widdowson, J. P., 45
Wilhowshe, C. J., 67
Wilkinson, H. W., 44, 46
Wilson, A. T., 46
Wilson, Armine, 35 fnt.
Wilson, G. S., 70, 97
Wilson, W. R., 67
Woodraffe, R. C. S., 67
Wortham, E. C., 44, 46

Y

Yow, M. D., 44

Z

Zabriskie, J. B., 44
Zahradnicky, J., 46
Zimmerman, R. A., 44

SUBJECT INDEX

A

Accu-Drop applicator, 92
Acellular organisms, 98
Acetic, 61
N-Acctyl-glucosamine, 58, 62
Achromobacter, 30
Acid and gas formation, 19-20
Acid formation, 19-20
Acinetobacter, 30
Acute glomeruloncphritis (AGN), 34
Acute-onset syndrome, 40-41
Acute rheumatic fever (ARF), 34
Adonitol, 22, 58, 60, 62-64
Aerotolerance, 61
Agar-agar, 79
Agar media, 86-87
Alcaligenes, 30
Alkalescens-Dispar group, 13
Alpha-hemolytic streptococci, 33
Alpha-methyl-glucoside, 14-15
American Heart Association, 34
Amygdalin, 58, 62
Amylose, 58, 62
Anaerobic bacteria associated with human disease
 Bacteroidaceae, 48-51
 Bacteroides bacteremia, 51-54
 Bacteroides fragilis ss fragilis, 47, 57-59
 biotyping of, 47-67
 micromethods for biotyping, 47, 54-56
 Propionibacterium acnes, 47, 57, 59-62
 endocarditis, 47, 62-64
Analytab Corporation, 55
Antibiotic prophylaxis, 34, 41
Antibiotic resistance genes, 17
Antimicrobial susceptibility testing of bacteria, 8
Antisera, 9, 35-36
API Lactobacillus 50 System, 47, 56-64
API 20 System, 54-55

Apyocyanogenic strains of *aeruginosa*, 27
Arabinose, 52, 56
D-Arabinose, 57-59, 62
L-Arabinose, 57-59, 61-62
Arbutin, 58, 62
Arginine, 22, 58
Arginine dihydrolase, 27
Autoimmunity, 34-35
Automation of biotyping, 71-75, 86-93, 95
 individual compartments, 87-89
 multiple inoculation, 89-92
 problems with, 92-93
 surface replication, 86-87
Autoplaque production, 28

B

Bacteremia, 51-54
Bacterial endocarditis, 57, 59
Bacteriocines, 9
Bacteriologic tests (*See* specific test)
Bacteriological Code, 70
The Bacteriology of Enterobacteriaceae, 70
Bacteriophage production, 17, 28
Bacteriophage-typing system, 36
Bacteroidaceae, biogrouping of, 48-51
Bacteroides, 52, 54
Bacteroides bacteremia, 51-54
Bacteroides biacutus, 50
Bacteroides corrodens, 53
Bacteroides fragilis, 48-51, 53-54
 ss distosonis, 49-54
 ss fragilis, 47, 51-53, 57-59
 ss ovatus, 49-50, 52, 54
 ss thetaiotaomicron, 50-54
 ss vulgatus, 50-54
Bacteroides incommunis, 48, 53
Bacteroides melanionagenicus, 53
 ss asaccharolyticus, 53
 ss intermedius, 53

Bacteroides oralis, 48-50, 53
 ss *elongatus*, 50
 ss *oralis*, 50
Bacteroides terebans, 48, 51, 53
Bacteroides trichoides, 50-51
Bacteroides variabilis, 48, 53
Bergey's Manual of Determinative Bacteriology, 15, 26, 29, 48-49, 101
Beta-hemolytic streptococci, 33, 38
Biliary tract infection, 43
Biochemical criteria, 13
Biochemical patterns of isolates, 10-11
Biochemical tests, 48-49
Biochemical variety (*See* Biovar)
Biogroup, 69-70 (*See also* Biotyping)
Bioserotype, 19
Bioserovar, 11, 28
Biotyping
 anaerobic bacteria associated with human disease, 47-67
 applications, 68-69
 automation, 71-75, 86-93, 95
 concept of, 6, 13
 defined, 4, 12, 47
 function, 5
 methods, 8
 micromethods, 47, 54-56
 nonfermentative gram-negative bacteria, 26-32
 perspective, 98-105
 program of, 21
 relationships in human disease or epidemiology, 68
 standardization, 68-97
 usefulness of, 48
Biotyping cube, 92-93
Biovars, 26-27, 29-30, 69-70
Blue green algae, 98
Butyric, 61

C

C substance, 33
Capillary inoculator, 90-91
Caproic, 61
Carbohydrate antigen, 33, 38-42
Cardiac valve prostheses, 57, 59
Cardiopulmonary bypass machinery, 60
Cardiovascular surgical procedures, 59-60
Catalase production, 60-61
Catenabacterium filamentosum, 51
D-Cellobiose, 58, 62
Center for Disease Control (CDC), 33-34, 38, 40, 54, 59, 65, 70
 Anaerobe Laboratory, 48-49, 60
Cephalothin, 65
CHO antigen, 38-42
Chronic respiratory infections, 27
Chronic valvular disease, 57
Citrate, 18, 22
Citrobacter, 13
Citrobacter diversus, 18
Citrobacter freundii, 13-14
Classification, 100-101
 enterobacteriaceae, 13-17
Clindamycin, 65
Cloramphenicol, 65
Clostridium ramosum, 51, 53
Colicins, 9, 17
Colonies of bacteria, 99
Color Standards, 71-72
Contamination sources, 57, 59-60
Cross reactivity, 9
 Iabc antibodies, 39
 streptococcal antigens, 35
Cultural tests, 48-49
Cystic fibrosis, 27

D

Dextrin, 58, 62
Dextrose, 61
A Dictionary of Microbial Taxonomic Usage, 12
Discrepancies in testing, factors causing, 93-94
Distilled water, 79-80
DNA hybridization, 15
 studies, 29-30
DNA relatedness values, 16-17, 99
Dulcitol, 19-20, 58, 62

E

Early-onset syndrome, 40-41
Electrophoretic mobility of soluble proteins of cells, 10
Endocarditis, 43, 47, 57, 59, 62-64

Subject Index

Enterobacter, 13
Enterobacter agglomerans, 14-17, 29, 31
Enterobacteriaceae, 9-10, 30
 biotypes of, 21-23
 biotyping in clinical laboratory, 12
 classification, 13-17
 clinical significance of specific biotypes, 18-19
 epidemiology, 19-23
 identification, 17-18
Enterococci, 43
Enzyme-linked immunosorbent assay (ELISA), 38
Epidemiological markers, 35-36, 40, 69
Erwinia, 15
Erwinia herbicola var. *ananas*, 16
Erwinia herbicola var. *herbicola*, 16
Erwinia stewartii, 16
Erwinia uredovora, 16
Erythritol, 58, 60, 62-64
Erythromycin, 65
Escherichia coli, 13, 17-19, 30-31
Esculin, 14-15, 52, 55-56, 58, 61-62
Eubacterium filamentosum, 51

F

F-factor, 17
Facultative anaerobes, 59
False positive, 80
Fermentation (*See* specific substrates)
Fingerprinting bacteria, 68
Fingerprinting in hospital epidemiology, 94
Flavobacterium, 30
Fluctuating biotypes, 77-78
Fluorescein production, 28
Fluorescent pseudomonads, 26-32
Fructose, 60, 63-64
D-Fructose, 58, 62
Fulminant sepsis-gangrene, 34
Fusobacterium, 52
Fusobacterium bacteremia, 54
Fusobacterium fusiforme, 53
Fusobacterium girans, 53
Fusobacterium mortiferum, 53
Fusobacterium necrophorum, 53
Fusobacterium necleatum, 53
Fusobacterium ridiculosum, 53

G

Galactose, 58, 60, 62-64
Gas formation, 16
Gas production in glucose broth, 52
Gas liquid chromatography (GLC), 48, 52, 62
 cell wall pyrolysis specimens, 10
Gastroenteritis, 18-19
Gastrointestinal tract, 57, 59
Gelatin, 52, 61
Genetic criteria, 13
Gentamicin, 65
Genus, 6-7
Glucose, 19-20, 52, 56, 58, 60, 62
D-Glucose, 58
Glycerol, 56, 58, 60-62
Glycogen, 58, 62
Green pigment production, 28
Group A streptococci, 34-38
Group B neonatal disease, 40-41
Group B streptococci, 38-42

H

H_2S production, 18, 21-22, 52, 55, 61
Hair follicles, 57
Hemolysis, 28, 61
Herbicola group, 15
Homo sapiens, 29
Hospital-acquired infections, 69
Host immunity, 38
Host tissue antigens, 35
Human errors, 80
Hyaluronic acid, 35 fnt.

I

Identification of *enterobacteriaceae*, 17-18
Identification of Enterobacteriaceae, 70
Identification of the Genera of Bacteria, 70
Identification of Medical Bacteria, 70
Impetigo, 34
Incorrect biotyping results, 80-81
Incubation time, 80-82
Indol, 14-16, 22, 50, 52, 55, 61
Infrared spectrophotometer, 10
Inoculum, influence of, 80-82
Inositol, 62

M-Inositol, 58
Internaitonal Code of Nomenclature of Bacteria, 69, 101
Intestinal tract infection, 43
Introduction to biotyping in clinical microbiology, 3-5
Inulin, 58, 62
Isobutyric, 61
Isocaproic, 61

K

KCN production, 18, 21-22
Klebsiella, 31
Klebsiella pneumoniae, 19

L

Laboratory contaminants, 57
Lactic, 61
Lactose, 14-15, 18, 22, 52, 56, 58, 61-62
Late-onset Group B disease, 41
Lecithinase, 14
Lincomycin, 65
Lombard-Dowell (LD) broth, 55-56
Low birth weight, 41
Lysine, 22

M

M-antibodies, 38
M protein antigens, 35-36 fnt., 38
M-typing reactions, 35-37
Malonate utilization test, 82-83
Maltose, 27, 52, 56, 58, 61-62
Mannitol, 19-20, 27, 50, 52, 56, 58, 61-62
Mannose, 56, 61
D-Mannose, 58, 62
Manual of Microbiological Methods, 70
Maternal genital colonization, 41
Matt colonies, 35 fnt.
Media, 70
 influence on biotyping results, 79-80
Melezitose, 62
D-Melezitose, 58
Melibiose, 14-15
D-Melibiose, 58, 62
Metabolic plasmids, 17-18
Metabolic products, 61

analysis, 48
Metastatic abscesses, 60
Methyl-D-glucoside, 58, 62
Methyl-D-mannoside, 58, 62
Methyl red, 22
Methyl red test
 end point determination, differences in, 86-87
 positive, 82-83
Methyl xyloside, 58, 62
Micromethods for biotyping, 47, 54-56
Milk, action in, 52, 61
Mini test dishes, 88
Minimal inhibitory concentration (MIC) in antibiotic susceptibility patterns, 64-65
Minitek Miniaturized Microorganism Differentiation System, 55-56
Minitek System (BBL), 55-56
Miscellaneous strains, 29
Moraxella, 30
Morphologic features, 48
Motility, 27, 61
Mouth, 57
Mucoidal colonies, 35 fnt.
Multineedle inoculator, 90-91
Multiple drug resistance, 17
Multiple inoculation, 89-92
Mutants, 27

N

Nasal cavity, 57
Nitrate, 14, 58, 62
Nitrate reduction, 16, 52, 56, 61
Nitrate to gas, 27
Nomenclature, 75-77
Nonfermentative gram-negative bacteria, biotyping of, 26-32
Non-hemolytic streptococci, 33
Nosocomial infections, 3, 7, 19, 22-23, 41
Nucleic acid hybridization, 10

O

102-peg test-tube rack, 89
Ornithine, 22
Ornithine decarboxylase, 14-15
Oropharynx, 59

P

Penicillin therapy, 65
  resistance to, 43
Peptone yeast extract glucose (PYG) broth, 52
Peritonitis, 43
Perspective of bacterial biotyping for clinical microbiology, 98-105
Phag lysis phenomenon, 8
Phage type, 11
Plasmid defined, 17
Plasmid mediated reaction, 18
Pneumonia cases, 19
Pre-reduced anaerobically sterilized (PRAS) media, 48-49
Principles of Bacteriology and Immunology, 70
Prokaryotic cellular organization, 98-99
Prolonged rupture of membranes prior to delivery, 41
Propionibacterium acnes, 47, 57, 59-62
  endocarditis, 47, 62-64
Propionic, 61
Propionic acid production, 60
Protective antibody, 35-36, 38
Protein antigens, 38-42
Proteus species, 13, 18, 31
Providencia species, 13, 18
Pseudomonas, 30
Pseudomonas aeruginosa, 9, 26-29
Pseudomonas aureofaciens, 26
Pseudomonas chloraphis, 26
Pseudomonas cichorii, 26
Pseudomonas fluorescens, 26-32
Pseudomonas putida, 26-27
Pseudomonas stringae, 26
Pyocin sensitivity, 28
Pyocins, 9
Pyocyanin production, 27

R

R-factor, 17
R proteins, 36 fnt.
Radioimmunoassay (RIA), 42
Raffinose, 14-15, 18
D-Raffinose, 58, 62
Refluxing biovars, 28
Reproducibility, 95

Resistance transfer factor (RTF), 17
Respiratory distress, 40
Results of biotyping
  incorrect, 80-81
  influence of media on, 79-80
  reporting of, 75-86
Rhamnose, 18, 27, 50, 56, 58, 61-62
Rhamnose-negative, 18-19
Rhamnose-positive, 14-15, 18-19
Ribose, 58, 62

S

Saccharolytic bacteroides, 48-59, 51
Salicin, 14-15, 56, 58, 61-62
Salmonella, 13, 17, 19
Salmonella cholerae-suis, 20
Salmonella enteritidis, 20
Salmonella serotypes, 19-20
Sebaceous glands, 57
Sepsis, 40
Septicemia outbreak, 16
Serologic procedures, 9
Serology, 13
Serotyping, 9, 11-13, 19
Serratia marcescens, 18
Serum opacity reaction (SOR)
  negative, 36-37
  positive, 36-37
Sex factors, 17
Sexual transmission of Group B disease, 41
Shigella, 13, 17-18
Shigella flexneri, 17, 19-20
Shigella sonnei, 17
Shigella typhimurium, 18
Significance of biotypes, 6-11
Skin infections, 34
Skin microorganisms, 57, 59
Sodium citrate, 58
Sorbitol, 58, 60, 62-64
L-Sorbose, 58, 62
Species, 6-7, 29
Species lumpers, 29
Species splitters 29
Standardization of biotyping, 68-97
Staphylococcus aureus phage type patterns, 11
Starch, 58, 61-62

Starr-Edwards ball-valve prostheses, 59
Steers replicator, 89-90
Strains, 7, 11
Streptococcal groups of clinical significance, 33-46
　Group A, 34-38
　Group B, 38-42
　Groups C, D, E, F, and G, 42-43
Streptococcal pharyngitis, 34
Streptococci, 9
Subjectivity in calling test positive or negative, 82, 84-86
Subspecies, 7, 11
Succinic, 61
Sucrose, 15, 22, 27, 56, 58, 61-62
Sulfadiazine sensitivity, 28
Surface-protein antigen, 35
Surface replication, 86-87

T

T agglutination, 35-37
T antigens, 36 fnt.
T-typing reactions, 36-37
Taxonomy, lack of agreement among experts in, 100-101
Temperature, 78-79
Test procedure
　description, 70
　influence, 82
　subjectivity in calling positive or negative, 82, 84-86
Test reproducibility, 95
Test tube holders, 89
Tetracycline, 65
Thioglycollate, 62
Throat infections, 34
Tight biochemical species, 29, 75
Time, influence of, 80-82
Trehalose, 14, 50, 56, 61-62

D-Trehalose, 58
True species, 29
Typticase Soy Broth, 62

U

Unity of biochemistry and biological order, 98-99
Urea degradation, 18, 21-22
Urease, 27, 52, 61
Urethral cultures of sexually active males, 41
Urinary tract infection, 19, 43

V

V-P, 22
Vaginal colonization, 41
Variability, 93-94
Variable species, 29
Virginia Polytechnic Institute (VPI), 48
　Anaerobe Manual, 49-50
Voges-Praskauer reactions, 16
Volatile acids, production of, 52

W

Whirlpool-associated skin rash, 28
Wild type strain, 76

X

Xylose, 14-15, 52, 56
D-Xylose, 57-59, 61-62
L-Xylose, 57-59, 62

Y

Yersinia, 13
Yersinia enterocolitica, 14-15, 18-19

QR
47
B56

DEC 10 1979